Public Perception of Drinking Water Source Protection: Who Wants What?

Blair E Nancarrow
Geoffrey J Syme

August 2010

DISCLAIMER

Water Research Australia Limited, is a not for profit company funded by the Australian water industry. (The name of Water Quality Research Australia (WQRA) was changed to Water Research Australia in July 2013)

Water Research Australia and individual contributors are not responsible for the outcomes of any actions taken on the basis of information in this research report, nor for any errors and omissions.

Water Research Australia and individual contributors disclaim all and any liability to any person in respect of anything, and the consequences of anything, done or omitted to be done by a person in reliance upon the whole or any part of this research report.

This research report does not purport to be a comprehensive statement and analysis of its subject matter, and if further expert advice is required, the services of a competent professional should be sought.

Location:

WaterRA Head Office
Level 3, 250 Victoria Square, Adelaide SA 5000

Postal Address:

GPO BOX 1751, Adelaide SA 5001

For more information about WQRA visit the website www.waterra.com.au

A methodology to incorporate community and multiple perspectives in the development of acceptable drinking water source protection policy in catchments facing recreation demands

Final Report 1023-09

ISBN 978-1-921732-04-1

Co-published by IWA Publishing
Alliance House, 12 Caxton Street, London SW1H 0QS, UK
Tel. +44 (0)20 7654 5500, Fax +44 (0)20 7654 5555
publications@iwap.co.uk
www.iwapublishing.com

ISBN: 9781780406183 (Paperback)
ISBN: 9781780406190 (eBook)

FOREWORD

Research Report Title: A Methodology to Incorporate Community and Multiple Perspectives in the Development of Acceptable drinking Water Source Protection Policy in Catchments Facing Recreation Demands

Project Leader: Blair Nancarrow and Geoff Syme

Research Nodes: Syme and Nancarrow Water

WQRA Project No. 1023

EXECUTIVE SUMMARY

This document reports the methodological development in the WQRA project *Public Perception of Source Protection and Its Relationship to Recreation and Water Treatment.* This project aimed to develop a nationally applicable methodology to assist in the formulation of defensible policy which provides for drinking water source protection while accounting for recreational needs in surface water catchments in Australia.

Climate change is reducing water resources in many parts of Australia. Increasingly, higher risk sources are either being developed, or considered for development as public drinking water sources. There is also rising demand from the community for additional recreation facilities and subsequent access to drinking water catchments, as they are predominantly close to urban areas, particularly to cities. These competing factors place substantial pressure on public water utilities to ensure the provision of safe drinking water. As the health risk profile increases, there is a need for additional barriers which may include new or upgraded water treatment facilities.

The key question being asked nationally is what is the relationship between the general public's requirement for safe drinking water and the demands of special interest groups to open up the drinking water catchments for recreational uses? What are the trade-offs between drinking water quality, cost of treatment, recreational requirements, and professional risk assessments. This project investigated public perceptions of source protection in relation to risks associated with public access for recreation in drinking water catchments as a case study in Western Australia (WA). This case study formed the basis for the methodological development.

The project was overseen by a national Steering Committee with representatives from WQRA and all Australian States and Territories except Tasmania. A local Working Group of key agency personnel in WA assisted with the ongoing application of the developing methodology to the WA context.

The developed methodology is reported here as well as providing the results and outcomes of the WA Case Study. Detailed results of the Case Study are provided in the Appendices, while result summaries illustrate aspects of the methodology in the main body of the report. The development of the methodology was underpinned by scientific and industry literature reviews which are provided here, and the methodology consists of three phases.

The first phase is **scoping** with the purpose of gaining *an understanding of the key issues* associated with the local context of drinking water catchment protection policy. This is done in a number of ways:

- understanding the issues from the perspectives of the *range of stakeholders and professionals;*
- understanding the relevant legislative framework that underpins drinking water source protection and recreational access to drinking water catchments; and
- determining cost implications of any changes to source protection and recreational access to drinking water catchments associated with the requirement for different levels of additional treatment.

The second phase is a **survey** which aims to investigate: knowledge of drinking water sources and acceptance of the current policy of catchment protection; risk tolerance; attitudes to uncertainty and potable water purity; aesthetic and landscape preferences; trust in water treatment, managers and science; the relationship between location and substitutability of other places for the drinking water catchments for recreation; lifestyle and recreational preferences; acceptability of alternative policy options and attitudes to cost and who should pay and demographic data that has been shown to influence attitudes to risk.

This phase of the project involves similar surveys of three different samples and builds on the findings of the Literature Review and the Scoping Phase. The three samples are: general community; members of formal recreation groups; and professionals associated with drinking water, catchment management and recreation. Apart from providing considerable useful information for involved agencies, key survey outcomes are used in the final phase of the methodology.

The aim of the final **risk-benefit assessment** phase of the project is to utilise the results of the survey, in particular the identification of acceptable and somewhat acceptable policies, and conduct a risk-benefit assessment to determine the optimal policy for implementation. The assessment involves the range of professional expertise as has already participated in the project to this phase.

In the original proposal it was thought that the risk assessment model developed by the CRC for Water Quality and Treatment would form the basis of the methodology chosen for this final phase of this project. However it soon became evident that, due to two factors that emerged in the preceding phases of the WA Case Study, an alternative method needed to be developed.

Firstly, while there was a consensus among the professionals that the current policy of catchment management had "worked" for over a hundred years, there was far from detailed agreement as to the science of how and why it had worked. In fact, the science of the adverse effects of recreation on water quality, while having some considerable support, was doubted by others who observed recreation occurring in water catchments elsewhere in Australia and overseas without any apparent ill effects. The queries about the actual reality of the risk applied to all three policies that emerged from the survey analysis for investigation.

Secondly, there were obvious concerns that there were significant benefits being foregone in restricting access for recreation in the catchments. There were clear indications that there was growing dissatisfaction with abilities to fully enjoy recreation activities of choice. As with the water quality science mentioned above, the extent of these benefits are difficult to precisely define.

Therefore, a risk-benefit methodology was developed which is underscored by a Multi-Attribute Utility approach and interpreted through a group technique of cross impact assessment. Cross impact matrices are developed for each policy option and are a simple method to identify the significant issues that will need to be addressed when planning for the implementation or introduction of a policy in the future.

CONTENTS

FIGURES

TABLES

1. INTRODUCTION

This document reports the methodological development in the WQRA project *Public Perception of Source Protection and Its Relationship to Recreation and Water Treatment.* This project aimed to develop a nationally applicable methodology to assist in the formulation of defensible policy which provides for drinking water source protection while accounting for recreational needs in surface water catchments in Australia.

Climate change is reducing water resources across Australia. Increasingly, higher risk sources are either being developed, or considered for development as public drinking water sources. There is also increasing demand from the community for additional recreation facilities and subsequent access to drinking water catchments, as they are predominantly close to urban areas, particularly to cities. These competing factors place substantial pressure on public water utilities to ensure the provision of safe drinking water. As the health risk profile increases, there is a need for additional barriers which may include new or upgraded water treatment facilities.

In the past, Melbourne Water has perhaps had the best catchment protection system, the envy of the world in terms of safe drinking water supply. Yet the demand for source augmentation in the light of increased population and reduced rainfall has meant that drinking water supplies are now being drawn from areas traditionally used for recreation and agricultural irrigation.

Perth has strong source protection mechanisms. Existing regulations prohibit the use of the water bodies for recreation and a 2km exclusion zone exists around reservoirs. However, recent experiences of conflict with recreators over the development of a potential new drinking water source in the south-west highlighted the need for the development of strategies to maintain the protection of existing and future drinking water sources. The 2010 Western Australian Parliamentary Inquiry into *Recreation Activities within Public Drinking Water Source Areas* challenged drinking water professionals and catchment managers to justify the current restrictive policy.

The key question being asked nationally is what is the relationship between the general public's requirement for safe drinking water and the demands of special interest groups to open up the drinking water catchments for recreational uses? What are the trade-offs between drinking water quality, cost of treatment, recreational requirements, and professional risk assessments. These trade-offs are being faced by the water industry and catchment protection authorities throughout Australia. The Sydney Catchment Authority is frequently under pressure to relax access restrictions to Sydney's drinking water catchments. Alternatively, the South East Queensland community is often critical of the water quality in Wivenhoe Dam which is open to recreation. With the need to augment supplies in the face of climate change, pressure may be put on traditional recreational areas with little or no substitutability and with the subsequent increase in risks to safe drinking water. Some of the risks can be partly alleviated by increased levels of water treatment, but would the public be prepared to pay for this?

This project investigated public perceptions of source protection in relation to risks associated with public access for recreation in drinking water catchments as a case study in Western Australia (WA). This case study formed the basis for the methodological development and had the following specific objectives.

- To determine the WA general public's and stakeholders' (including professionals') acceptance of a range of policy options in the management and protection of drinking water catchments, and the influence on acceptance of knowledge of catchment protection; knowledge of potential risks and risk tolerance; attitudes to uncertainty and potable water purity; aesthetic and landscape preferences; trust in water treatment, catchment managers and science; the relationship between location, size of resource and substitutability; lifestyle and recreational preferences; and attitudes to cost and who should pay for any additional treatment by:
    - conducting a scoping phase in WA which includes:
        - a literature review of relevant international and national water industry, medical, environmental and recreational research;
        - a series of interviews and focus groups with relevant stakeholders and members of the public to establish the parameters of the above variables;
        - the examination of the relationship between current State catchment protection and various agencies recreational policies to establish conflicts and synergies; and
        - assessment of the costs of additional treatments in the case of multiple catchment use risks to quantify public cost burden.

- planning and carrying out public and stakeholder surveys on the determinants of public perceptions of catchment protection and acceptable protection policies in relation to the purity and treatment of the water supply and recreation on the catchment.
- undertaking a professional risk assessment of the range of acceptable policies determined in the survey in relation to dual use of catchments and the levels of required treatment.

- To provide justification and recommendations for catchment management policy, and also any requirements for community education of the issues.
- To document the outcomes and methodology as Stage 1 of the project for testing and refinement in interstate case study(ies) with the final product of a nationally applicable methodology.

A Steering Committee with national representation oversaw and advised on the conduct and Milestone reporting of each stage of the project. The Steering Committee had representation from the following organisations.

- Water Corporation - WA (Chair)
- Water Quality Research Australia
- National Water Commission
- Urban Water Security Research Alliance - Queensland
- Sydney Catchment Authority
- Melbourne Water
- SA Water
- Power and Water – Northern Territory

A local Working Group was formed to provide WA specific advice and information. The Working Group consisted of nine members with a range of relevant expertise from the Water Corporation, the Department of Water and the Department of Health. The Department of Sport and Recreation was also invited to provide a representative to contribute to the Working Group but the invitation was declined due to heavy work commitments.

This report details the methodology that was developed as well as providing the results and outcomes of the WA Case Study. Results of the Case Study are provided in the Appendices, while summaries illustrate aspects of the methodology in the main body of the report. The development of the methodology was underpinned by scientific and industry literature reviews which are provided here, and the methodology consisted of three phases.

- Scoping
- Survey
- Risk/benefit assessment of acceptable options/policies determined in the survey phase.

2. LITERATURE REVIEW AND RESEARCH FRAMEWORK

2.1 The Social Context of Drinking Water Catchment Protection

2.1.1 Background

The rivalry between the protection of drinking water catchments for water supply purposes and recreational demands has been increasing over the decades, and will continue to increase especially in areas where climate change forecasts have predicted a continuing drying trend.

While policies on drinking water catchment protection vary from State to State, largely because of legacy issues, the generally preferred response where possible is to avoid recreation and other land uses that may increase the risk of contamination of the water supply. In the case in Western Australia, this has meant a policy of an exclusion zone of two kilometres immediately adjacent to the water bodies in the drinking water catchment. A wider area has limited access for recreational activities. This area has been planned based on prevention considerations and the practicality of legacy issues. In short, the philosophy is that prevention is better than cure, that drinking water catchments have an important role in maintaining the diversity of supply should technology fail or other sources become contaminated or drinking water quality incidents occur. Should treatment be required because of human activity there would be the need for multiple barriers regardless of the level of activity as even apparently small outbreaks of pathogens have the potential to cause significant outcomes (WSAA, 2008).

Any significant activity, especially in the exclusion zones, would thus move the tried and true "prevention" model to one of "risk management" with its attendant problems of technological and human failure. Human failure is thought to occur in the normal operation of technological systems, but also in the behaviour of those on catchments some of whom will never adhere to community rules, such as correct disposal of faecal material, fire prevention and operating in designated areas. Finally, the public outcry that would occur should drinking water become contaminated is likely to be enormous in cities such as Perth after more than a century of safe water supply which is now taken as a given. The fallout from such an event is likely to be severe both for the water service provider and the State Government as seen in the reaction to the Sydney incident in the twentieth century (WSAA, 2008).

Finally, there are wider issues that need consideration in catchment protection and these include the positive externalities of human exclusion such as the potential to protect a wider range of biodiversity and landscape amenity values.

Nevertheless, the benefits of outdoor, bush or wilderness based recreation are well documented. There is increasing demand generally for such recreation with rising populations. As a result, there is heightened potential conflict between passive and active recreators as they attempt to use the same or adjacent areas. Water based recreation is a substantial component of such recreation demand.

This mounting demand is increasingly problematical for recreation planners as often they have no control over land use planning. Therefore, recreation needs are one of many considerations in the planning process. Unlike water supply, there is no prior protection of catchment based recreational "production" and opportunities have to be created. This has led to some groups such as water skiers, who cannot use the ocean, feeling disenfranchised, and others such as trail bike riders with limited designated sites operating on an ad hoc basis often to the environmental and amenity concerns of others. Recreation planners also consider that they are singled out as a land use, especially in the outer catchment where agriculture and industry also may have "legacy" incursions. Because of this likely increase in demand, the pressures for evidence-based protection of drinking water catchments (particularly the outer catchments) will require credible justification and be seen to reflect the overall public interest.

For recreation planners, the challenges associated with accommodating recreation in Western Australia generally will be greatly compounded by climate change as increased drying occurs. There will be increased evaporation on water bodies, both natural and artificial, water levels will drop in recreational dams thus reducing their capacity. Finally, the WA Department of Sport and Recreation's (2006-2010) strategy (DSR, 2006) suggests that, in the future, there will be significant indirect effects of increasing water costs presumably affecting artificially maintained water bodies and local pools. It will also affect the "emergency" supply that some recreation lakes provide for rural areas in wider Western Australia.

Finally, some recreators point to the difficulty in science "proving" the link between catchment protection and health consequences using existing drinking water catchments which accommodate recreation as examples, as well as the tenuousness of examples of water-based health incidents linked to catchment management. Examples of international practice are given as demonstrating that treatment can be effective on multiple use drinking water catchments. It has been stated that the limited examples of health outbreaks are largely from swimming pools or in one case a public fountain. (Hughes et al., 2006).

However, the movement of the Wachusett Reservoir from a largely unprotected to a protected status (Steinberg and Clark, 1999) tended to show that "recreational activities on watershed land as having a moderate impact on reservoir quality (less significant than septic tanks, urban runoff or waste facilities but more significant than road salting, barnyards or erosion)". Further, while there are knowledge gaps, Ferguson et al. (2003) state that while it is clear that "transport of pathogens overland in surface runoff is clearly responsible for event-related increases in the concentrations of in-stream water borne pathogens in many watersheds, there are still several knowledge gaps in understanding the catchment processes in this regard". Nevertheless, in developing a pathogen movement process, Ferguson et al. (2007) noted that "the model identifies that pathogen excretion rates from animals and humans and manure mobilisation rates are significant factors determining the output of the model and thus clearly warrant further investigation". There is obviously sufficient evidence for continuing to develop our understanding of pathogen processes in the catchment context.

Despite these findings there are some who would suggest that the philosophy of protection is a flawed approach and that a "freedom to roam" policy should be adopted which is modified only if problems are identified. Recreation is rightly seen as only one of a number of catchment activities. Effective remedial action can be taken for any recreation influences on water quality. Concern is also expressed that the catchment area (especially the wider limited access zone) is too large at present in Western Australia (Hughes et al., 2008).

Fortunately, while there is already some dissatisfaction by pockets of recreators, the conflict between drinking water catchment protection and recreation has been largely manageable to this point. But it is important that community perceptions of the issues surrounding catchment protection including its philosophy, risk management, public good aspects of water supply, the individual rights of both water consumers and recreators, attitudes towards prevention or avoidance as a risk management strategy rather than active risk management, the nature of scientific certainty, risks of concern, the perceived reversibility of both recreation and drinking water catchment decisions, potential conflict between recreators and the concerns for "at risk" recreational pursuits. Without such knowledge strategic planning, either for drinking water quality protection or for recreational planning, will be exceedingly difficult. From the recreational viewpoint many of these perceptions will be relevant to the sustainable management of catchment-based recreation generally.

In this review, we examine the issues both theoretical and methodological that surround the above perceptions with a view to establishing a general social research agenda and specifically to guide the development of the methodology in this study. The review is not intended to cover the background issues of health and catchment management which have already been dealt with elsewhere. Nevertheless, we are motivated by the findings of Bruce, (2006a, b) whose work provides an excellent background to the social "problem". People who felt as though they were to lose their existing recreation activities were strongly opposed to the catchment protection that would be afforded to drinking water designation. The general public however wished to be assured that their water supply purity was guaranteed and thus supported catchment protection. The positions are not necessarily directly opposed, but will demand careful social investigation of catchment allocation and management decisions in the light of future population growth. This needs to be conducted both at the stakeholder and general public levels.

Here, while the current case study is the South West Australian catchment situation, the issues and methodologies discussed are equally relevant to drinking water catchment protection generally.

2.1.2 The WA Environmental Context

The protected catchments of most interest to the community are those shown in Figure 1. It can be seen that these catchments provide significant potential for outdoor recreation close to the city for many. Accurate figures of current usage of the outer catchments are difficult to obtain apart, perhaps, from those of organised groups. It appears that, from environmental scrutiny and reports by rangers, there is significant informal use for a variety of activities (e.g. bushwalking, trail bikes and fishing) on the outer catchment and also to some extent on the Reservoir Protection Zones (RPZ) themselves. Thus, pressures exist now and can be predicted to rise as Perth's population grows.

The area of land prescribed as exclusion zones in Western Australia consists of an inner RPZ of 2kms (outlined in red in Figure 1). There are also outer catchment areas of significantly larger scope which are available for restricted recreation. The desire for greater use of drinking water catchment areas (especially the outer zones) is mounting. This increase in demand, combined with the considerable recreator dissent when the Logue Brook was proposed to be changed from recreation uses to public water supply, indicates that any decision to allow greater recreation in the current drinking water catchments may well be politically irreversible.

2.1.3 Measuring and Understanding Attitudes towards Risk and Environmental Protection

In this section, we examine the key issues from a social psychological perspective. The measurement of attitudes will only be helpful if it creates a robust understanding of the reasons for the difference of opinion between people on the issue and ways forward for both water and recreational planners. For example, it is highly probable that those who are involved in the promotion and administration of sports, or those who are merely avid participants will have well formed views based on the experience of years, and an understanding of the impacts of the sport. Part time recreators or casual individuals who enjoy the recreation occasionally may have less definite views and these may have a different psychological basis from those who are group members and have social as well as recreational outcomes from their participation. Finally, the wider public, who have only sporadic interest in outdoor recreation, may have more general values-based attitudes which relate to issues of environmental protection and amenity, and health issues associated with catchment protection and management. There are multiple aspects of perception, attitudinal and preference measurement and their relationship to behaviour. These are explored below, and will provide the basis for addressing the issue of catchment protection and recreational demand in a wider context.

In exploring some of these options, the intention is to provide a potential "toolbox" for the conduct of this WQRA project. Ideally, whichever methods are feasible, they can be implemented and analysed to create a predictive model for support of catchment protection policy that can enable careful attention to the needs of differing groups and the general public (e.g. Kline, 2005).

2.1.3.1 Accounting for Attitude Strength and Polarisation

Most attitudinal investigations measure the positive or negative responses to a statement (Damodar, 1995). This is also the case for this investigation. Nevertheless, attitudes have a number of structural characteristics that are important to understand if we are to predict behaviour or preferences from them. Attitudes may be well formed, important, strongly or weakly expressed, polarised within the community, moderate or extreme. These characteristics are sometimes collectively called "centrality" and are not always measured, but seemed to have great relevance for this investigation (see Petty and Krosnik, 1995 for a review of this issue). The issue of polarisation is potentially important here because of the contrast in perceptions of how risk should be dealt with either from the preventative or the "intervention when necessary" approach.

Weak attitudes, for example are much less likely to be predictive than strong or central attitudes as they are sometimes constructed "on the spot" to meet an ephemeral circumstance such as someone telephoning to ask them what they think. Stronger or more central attitudes though are postulated to be more accessible and they connect to behaviours and preferences more readily. Thus it was hypothesised that the more central or more polarised the attitudes were in the community the more they would predict preferences (Krosnick, 1988).

Having given a very succinct description of the attitudinal strength literature, it is important to note that there is some disagreement about how this variable should be measured. Many dimensions of attitude strength have been measured. It is suggested for this issue, the five dimensions which seem to most often recur in the literature are highly relevant: perceived centrality (or relationship to other attitudes); degree of interest in the topic; the intensity with which the attitude is held; the importance of the attitude to the respondent; and the certainty with which the attitude is expressed.

Attitudinal polarisation is also important to understand (e.g. Damodar, 1995). The attitudes, knowledge and beliefs associated with strong alternative positions on policies need to be examined carefully in regard to information dissemination and policy and management evaluation.

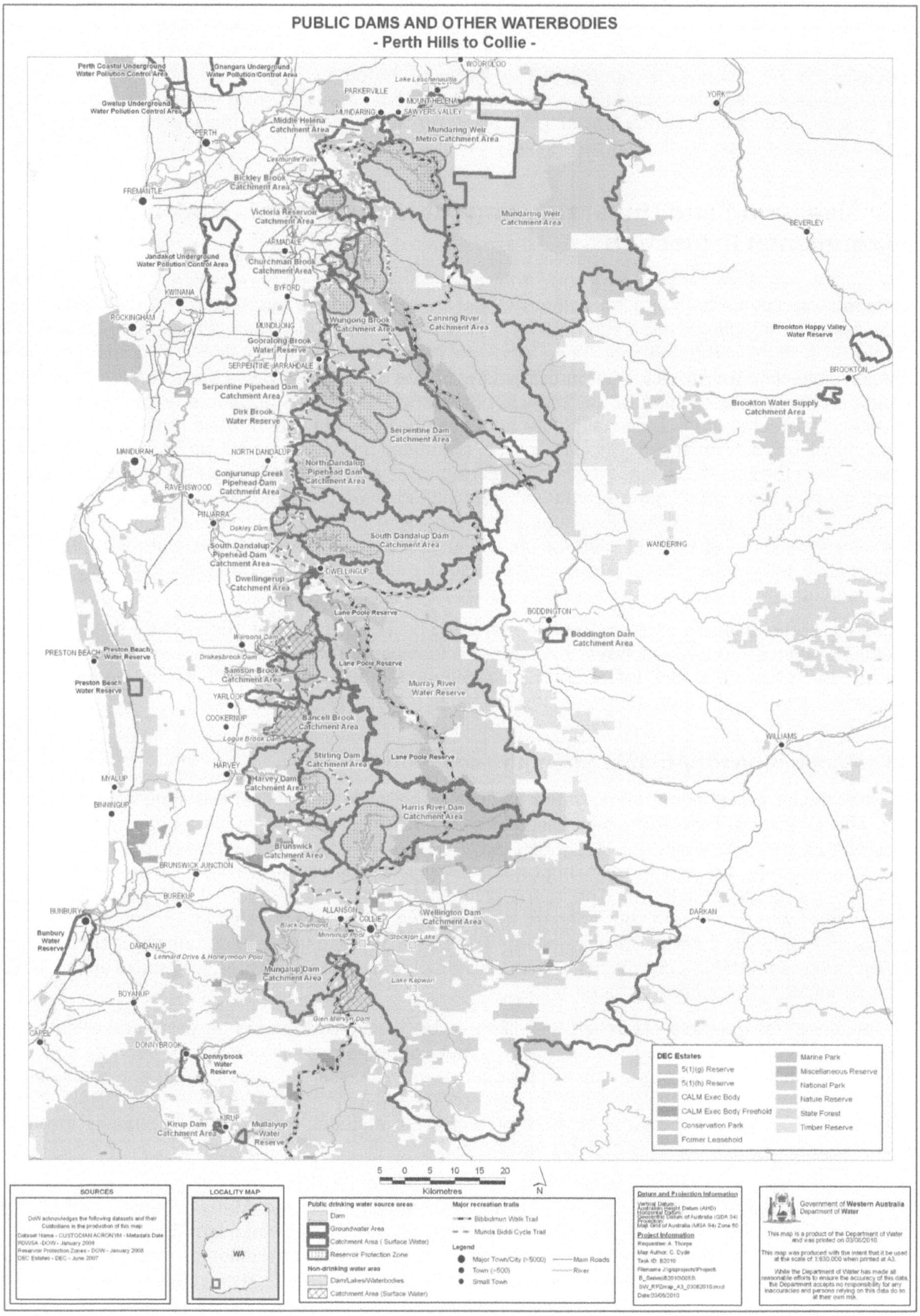

Figure 1. The WA Context (Source: Department of Water, WA)

2.1.3.2 Thresholds

As stated in terms of the polarisation of attitudes, catchment protection can be seen as a polarising issue for the public. The issue can be seen as one of opposing philosophies that of preventive rules or thresholds (exclusion) or that of risk management. The first relates to rules and the second to using a risk minimisation or multiple barrier approach to preventing water quality "accidents" in more open catchments.

The public perception can be a threshold one - which end should we start from, prevention or risk management, or are tradeoffs appropriate? There are many possible approaches to identifying and defining such issues. But there are two methods which can be applied to the catchment priorities.

The first approach (Connerley, 1986) examines factors which cause large shifts in community preferences when aspects of policy are introduced and taken away. Careful examination of why particular consequences of land or water use change suddenly can then be identified and investigated. For some issues, there are no key thresholds, but for others there are key outcomes that will not be tolerated by the community. For example, in the study of a potential industrial development in the Murray Shire there was a strong support of approximately eighty percent for the development, but this fell to fifteen percent if there was any credible threat to the groundwater quality (Eaton and Syme, 1989). Clearly further investigation was required of this outcome if community sanction for the development was to be gained. Groundwater quality in this case was clearly a threshold or pivotal issue. If prevention is a threshold issue for many in the community, then risk assessment becomes irrelevant to their decision making.

The second approach to thresholds that is relevant to the drinking water catchment protection issue arises from Sherif's (1961) social judgment theory. Sherif noted that when judgements of support for policies or socially relevant objects are made they are often measured on a single scale and with a single rating score. Social judgement theory suggests though, that in many cases where there are a variety of possibilities, one option may be the most preferred on a rating scale but that others may also be acceptable or rated favourably by the individual. Similarly, there may be some options which are not preferred and some that are unacceptable. Using this theory, it is possible to derive simple methods which measure both the latitude of acceptance of individual's policy preference (i.e. one's most preferred approach and others that may be acceptable) and one's latitude of rejection (ie one's least favourite option and others that are unacceptable). Perceptions and preferences can then be interpreted in terms of acceptable bands of options and generally unacceptable bands of options. The mapping of these latitudes can provide justification in social terms while providing some flexibility of examining different options. In the water resources area, this approach has been very successfully applied to water restrictions in the Perth metropolitan area (Williams and Syme, 1989).

2.1.4 Possible Determinants of Catchment Protection and Related Values, Attitudes and Preferences

2.1.4.1 Values and Ethics

There has been some interest in post-material values and ethics and their relationship to environmental management issues in recent years (e.g. Dietz et al., 2007; Lumley and Hercock, 2000; Syme, Nancarrow and McCreddin, 1999) and there have been some statistically significant relationships demonstrated. In particular, in Western Australia, Lumley and Hercock (2000) have demonstrated the relationship between these variables and voting patterns in state elections. Karpiak and Baril, (2008) have shown that people who are guided by moral principles also demonstrate more commitment to environmental issues. Ethical and fairness issues were also found to be highly significant in community trade-off preferences for the sustainable management of groundwater resources in northern New South Wales (Syme, Nancarrow and McCreddin, 1999). These variables are also likely to be an important determinant of attitudes towards catchment protection policies whether in relation to drinking water or for ecosystem preservation.

2.1.4.2 Trust in Government and Scientists

The issue of water quality whether in relation to potable wastewater recycling or water contamination in general has become one of high political and academic debate over the past few years. Willingness to pay for environmental protection has decreased over time as this issue enters into one of central concern to the community. Ivanova and Tranter (2008) believe that the routine nature of environmental issues has played a large role in diminishing this willingness to pay. When salient environmental problems arise however, Ungar (2000) states that the quantity and complexity of knowledge about issues such as water quality has left the broader community with the question of 'which expert do I think I can trust?'. Ross (2005) found that the major predictor of trust in the Water Corporation to provide safe drinking water was whether people identified with the Corporation as part of the Perth community. This "social bond" was significantly related to lesser belief in something going wrong with the water quality, and a greater belief of fair dealings by the Water Corporation.

Often to deal with issues in the face of complex science, the community uses trust as a summary judgement to underpin their views of the desirability or undesirability of policies in relation to contentious water issues with risk connotations (e.g. Nancarrow et al., 2009). For example in the case of acceptance of potable reuse of wastewater, trust in the competence of the water managers predicted risk judgments of the recycling science and technology as well as the risk that the operators were likely to make "mistakes". This issues needs to be more fully understood in relation to drinking water catchment protection.

2.1.4.3 Risk Perception

The source protection issue is a good example of a risk which may be represented in the community's mind as a high consequence, low probability event which is typical of many issues in water resources management (Syme, 1995; Sullivan, 1998). Extreme flooding, dam safety management and extreme precipitation events are in the same category. In the case of catchment protection, contamination events can be extremely serious in cities and can be the cause of significant outrage and a long term erosion of trust as has been demonstrated in the Sydney incident some years ago. This concern about extreme events is also exacerbated by the occurrence of climate change in Australia and this will be incorporated as an aspect of uncertainty in this context. This has been responded to in earlier research by the CRC for Water Quality and Treatment at an "expert" level (Miller et al., 2006).

Classically in these types of risk there are three significant pillars of risk acceptance. These (affectionately known to some as **TLC**) are current **T**rust, conviction that there is a clear **L**iability that can be relied on, and that there has been reasonably informed **C**onsent from the community. Trust has been summarised above. Liability has been more clear in recent years, and consent is the topic of this WQRA project. It is interesting that the three interact in the public mind. A previous study (Syme and Bishop, 1992) in relation to community perceptions included public considerations of the penalty that should accrue to the dam builders and operators should the dam fail. There were a number of scenarios presented which varied in the level of community involvement in setting the dam safety standard from none to conduct of a public interaction program. The penalty allotted by the community to the engineers was significantly higher when no community involvement had been facilitated compared with that when they had been consulted. The non consulting agency was awarded a significant jail penalty. Thus consent is related to liability.

At the level of the current project however there are a wide variety of conceptualisations of risk available (e.g. Bammer and Smithson, 2008). However it is suggested that in this case several dimensions of personal perceptions of risk need to be investigated. These are the perception that personal health is at risk from poor catchment management and uncertainty aspects of the management of the catchment to deliver healthy water. These uncertainty aspects are the perceived likelihood that a contamination incident will occur, the perceived seriousness of the consequences if contamination does occur and the perceived control that the government can exert to prevent contamination from the appropriate execution of catchment management programs or policies (Nancarrow et al., 2009).

The likelihood of an incident occurring can be conceptualised as the individual's belief that that society is able to organise itself through its institutions to effectively deal with an issue. In this issue it may be exemplified in effective operation of clean up technology or in the ability of our institutions to provide effective deterrence of undesirable behaviours in the catchment (Cilimburg et al. (2000). Focus group discussions to scope this study reveal that many do not consider there is sufficient policing of the current RPZs and there is poor use of voluntary resources to add to catchment security. Informal discussions with both enforcing public servants and the public frequently expose a dichotomy in natural resources management. Public servants feel they are in control and the public feel they are not, due to lack of resources.

A further difference arises with the perception that industry impacts can be controlled on the catchment. Public servants are more optimistic than the public on these matters.

Therefore, there is a need to examine potential formal and informal institutional development to promote feelings of social efficacy (Ivey et al., 2006). If perceptions of social efficacy are high, whether through our organisation or technological innovation, it is assumed that support for catchment management policies will be maintained. However, this concept has been relatively poorly researched at an international social psychological level.

2.1.4.4 Knowledge

Knowledge of an environmental issue is always thought to be an important factor in gaining support for environmentally responsible behaviour, whether from responsible recreation or environmental water quality protection viewpoints. This relationship has been assumed to occur in many environmentally relevant problems. Hence the perennial call for education programs.

In fact the relationship between education or knowledge and behaviours of potential significance to catchment behaviour such as acceptance of waters of appropriate quality or acceptance of personally inconveniencing climate change policies can often provide confusing or contradictory results. For the water quality example, knowledge does not seem to be an important variable in the acceptance of recycled wastewater for potable use (Nancarrow et al., 2009) but it is a highly significant variable in driving the acceptance of climate change policy (Syme and Nancarrow, 2009). Despite this contrast Bamberg and Moser, (2007) in a meta-analysis of twenty years of environmental behaviour, report that overall "problem awareness" which presumably has some relationship to knowledge is a significant indirect cause of environmentally responsible behaviour operating through a number of personal values and attitudes.

What significance knowledge of catchment processes in general, or recreational or water treatment technologies specifically, may have on policy preferences for protected catchments is not known at this point. Indeed the general knowledge in relation to such issues is an unknown quantity. What is certain however, is that any individual will have two components to their knowledge. These are actual knowledge about catchment processes and perceived knowledge of catchment processes. The latter relates to how knowledgeable people think they are (Syme and Nancarrow, 2009). The two may only correlate moderately and differentially affect attitudes towards catchment protection both in the context of recreation and other land uses.

Finally, in communicating catchment protection policy there is frequent reference to adhering to national and international drinking water standards and guidelines. Presumably this is meant to reassure the public that appropriate health protection measures have been taken. Recent research on public knowledge and perceptions of drinking water standards in the USA however shows that there was considerable public scepticism of such standards and their effectiveness (Johnson, 2008). Whether these reservations exist in Australia, and if so, are based on knowledge or public perception of risk needs to be further investigated in the catchment protection context.

2.1.4.5 Substitutability, Site Attachment and Specialisation

Concerns for protection of drinking water catchments from recreation relate especially in Western Australia to the fact that many of the protected catchments are close to the city and that equivalent experiences cannot be gained elsewhere. Further, for water based recreators who desire fresh water environments, open expanses of water are increasingly scarce, especially in the light of climate change in the South West of Australia. In addition the "outer" protected catchment is relatively large in area and contain a diversity of environment which is an important component of the recreational experience for many (Bestard and Font, 2009). Tanner et al. (2008) in examining visitor patterns to four protected areas in the USA have shown, for example, that while there were values assigned by the visitors that were common to all areas, there were some significant values that were unique to particular locations. "Understanding how and why assigned values vary by protected area is likely to be useful as managers confront complex and conflict laden issues".

The issue of the analysis of substitutability therefore is an important one in the social context especially from the recreation viewpoint. The concept of substitutability of recreational experiences has changed over time from the sole consideration of the experience of the leisure activity itself to a formulation which includes environmental features.

"The term *recreation substitutability* refers to the interchangeability of recreation experiences such that the acceptably equivalent outcomes may be achieved by varying one or more of the following: the timing of the experience, the means of gaining access, the setting and the activity" (Brunson and Shelby, 1993).

While the achievement of substitutability may seem to be relatively easy in the natural setting because of the relatively large areas, it is apparent that there are a number of constraining factors that need to be considered. The two most significant are site attachment and recreational specialisation.

Site Attachment

Site attachment is an increasingly studied phenomenon the understanding of which probably lies in the 'mere exposure theory' (Fink, 1989). Simply stated, the more one is exposed to an environment which provides an enjoyable experience the more one becomes attached to it. While this early work was developed in relatively simple experimental settings, it is evident in the natural environment including lake settings (Jorgensen and Stedman, 2006), and in the context of leisure (Eisenhauer et al., 2000). This is most clearly demonstrated in the repeat visitation to the same location for holidays and within the context of a wide variety of activities. From the point of view of catchment planning, ithe reversal of catchments from leisure to drinking water purposes creates a sense of "prior rights" or legitimate ownership of an area. This phenomenon can occur even in quite remote areas, for example on pastoral properties near the Ningaloo Reef where newcomers are chased from camping locations by those who have used those sites over a number of years.

Such "ownership" can then lead to the perception of the owners that they have an "endowment" which is held in very high value in comparison with other sites. Taking away of such an endowment from recreators is then likely to be seen as a personal loss of "rights" which according to Kahneman and Tversky, (1979) and others, will be values up to 10 times more than any new site offered. This may be the underlying cause of much of the conflict surrounding retrieval of recreation catchments.

Recreational Specialisation

The need for catchment based recreation will vary widely. Some people will wish to pursue passive recreation, others active bushwalking, rogaining or orienteering and others seek the conduit for motorised or other vehicular recreational transport. Some people will join groups and others will sample a wider range of recreational pursuits sporadically. Most of the discussion between government agencies and recreators as to appropriate policy tends to occur between organised groups and their representatives. These representatives tend to specialise and take their activity and its promotion 'seriously'.

The concept of serious leisure was introduced by Stebbins(1992). It was defined as "the systematic pursuit of an amateur, a hobbyist, or volunteer activity that participants find so substantial and interesting that, in the typical case, they launch themselves on a career centred on acquiring and expressing its specialist skills, knowledge and experience" (p3). Stebbins identified six criteria by which serious recreators could be distinguished, all of which are readily discriminated the serious from the more casual leisure. There have been a number of developments on this early work. Perhaps the most significant for this discussion was that all of the criteria were found to be related positively to past experiences and centrality to lifestyle (Tsaur and Liang, 2008). Thus in attitudinal terms such individuals are most likely to have strong attitudes, a strong identity with place and infer the greatest 'endowment'.

The centrality or importance of leisure is of interest to catchment planners as the more important that attributes of an activity are perceived to be, the smaller number of substitutes are perceived to be available for it (presumably including location) (Manfredo and Anderson, 1987). Sorice et al., (2009) in a marine setting have also found that committed divers were the least favourable of three groupings towards trip profiles with greater management restrictions. Nevertheless, Tucker et al. (2006) also found a positive relationship between site attachment and identification with river based leisure activities and intentions to pay for environmental improvement.

Clearly the issue of 'serious' recreation needs to be further explored in the context of drinking water catchment protection.

2.1.4.6 Equity Issues

As with all increasingly scarce resources with a complex mixture of private and public good characteristics, there will be a number of problems involved in assessing how allocation can be achieved in a fair manner. Social psychological studies of equity and fairness have tended to show that community views of allocation decisions are likely to hinge on three interacting factors: self interest; efficient uses of resources; and fair allocation of them (Wilke, 1991). The first factor requires a needs assessment as to what approach to protecting drinking water catchment would be preferred to maximise the individual's own utility. The second refers to the best outcome to obtaining the most benefit possible from the available resource. This is akin to traditional economic analysis. The third dimension of fairness however includes, as stated earlier, lay ethics and pro-social motivations (Syme et al., 1999) that are interpreted to ensure that sharing occurs in a way that benefits the community as a whole (or in old fashioned terms, the public interest is met).

One of the key issues in relation to this project will be "who pays" either from the viewpoint of recreation or catchment protection. Alternatively there will be discussions about who are the beneficiaries and to what degree.

2.1.5 Policy Evaluation: Assessing Benefits and Costs and Preferences

Traditional benefit-cost analyses can be simple or more complex and initially were conceived as dollar comparisons of attributes of a policy or management for goods or services that could be traded on the market. This can be problematic for issues such as those relevant to catchment protection. Many of the goods are intangible and can't be traded on the open market. From the recreational viewpoint these include the feeling of wellbeing, enjoyment of aesthetic beauty, social connection, the excitement of taking risks and even spirituality. From the environmental protection viewpoint there are issues of the value of the existence of species or biodiversity, concern for the next generation, and the enhancement of the moral connection to the environment. Many intangible benefits are shared between the recreation and environmental protection. On occasions, depending on the recreation type, they can conflict both between different recreation interests and between conservationists and recreators.

In this section we will briefly examine alternatives for measuring the value of the non-market values which are pertinent to catchment protection. These include methods to endeavour to convert measured human values and preferences into dollar values so that they can be incorporated into the basic benefit-cost analysis, and alternative measures that examine more general utility or values preferences for alternative policies.

2.1.5.1 Subjective Measures of Dollar Values of Non-market Goods

These measures use attitudinal scaling or contingent preferences to assess the potential increase of utility from changes in catchment management policies. The externalities, or effects on values other than that for which the focus of the policy changes (e.g. effects on ecological values from submersion of catchment areas for water supply) can also be assessed.

Such exercises have created significant philosophical debate (e.g. Ackerman and Heinzerling, 2004) by those who object to the utilitarian view of such an approach. There is also significant empirical work which suggests that the motives behind stated willingness to pay (WTP) are not appropriate for benefit-cost analysis. This is because the motivations attributed to the WTP judgement are ethical and socially constructed rather than economically motivated (e.g. Spash et al., 2009, Kahneman et al., 1999). In the discussion above on equity, it would seem that while economic efficiency is an important component of decision making there are others which also require consideration. In short, a dollar value may be only one input into the decision making process.

In what follows three of the most frequently used techniques, contingent valuation, contingent valuation and choice modelling are briefly examined. All have been applied to water resources problems (Loomis, 2000).

Notwithstanding this issue there is now a substantial literature on the reliability and validity of contingent valuation studies (what are you prepared to pay for a particular policy scenario or outcome [open ended] or would you be prepared to pay $x? [dichotomous choice] (see Mitchell and Carson, 1989). A second popular method is contingent ranking. In contingent ranking studies respondents are asked to rank or rate a series of 'product profiles' that describe products (or policies) and product attributes can then be used to estimate the marginal utility of each attribute. Since price is usually one of the attributes it is possible to rescale the ranking index in monetary terms. (e.g. Foster and Mourato, 2000). Choice modelling is a third approach to stated preference valuation (Rolfe and Bennett, 2006). It has the advantage of being able to deal with a larger number of attributes than contingent valuation studies through a series of experimentally designed choices in which the values of the attributes of the policy can be varied with choices between them guided by price. It has the ability to deal with potentially irreversible environmental choices for example clearing land for productive use (e.g. Mallawaarachchi et al., 2001). Like contingent valuation, there is a burgeoning literature relating to aspects of validity and reliability of the methodology such as choice range and payment vehicle effects.

There are relatively few studies which compare the dollar values between the differing techniques though a comparison between contingent valuation and contingent ranking studies. Bateman et al. (2006) has shown that in the case of urban river water quality improvements, contingent ranking led to a significantly higher dollar value than contingent valuation. A further issue which would need to be investigated further in the catchment context is the finding in road safety that there is little correlation between a respondent's precautionary behaviour, and stated willingness to pay for protection (Svensson, 2009; see also Wong et al., 2003).

The large number of protest votes attending economic valuation studies of this nature also needs to be considered. These can often be quite substantial in terms of proportions of respondents and there is some controversy about their interpretation (e.g. Jorgensen and Syme, 2000; Morrison et al., 2000).

Thus care needs to be taken in dollar valuation studies and their purpose has to be defined, especially in allocation decisions where it is likely that social and ethical concerns will predominate. A good guide to implementing such studies is provided in Bateman et al., (2002).

2.1.5.2 Non Dollar Based Assessment Tools

There are a number of decision ranking and rating approaches to assist the evaluation of different approaches to catchment protection which can be employed either in planning workshops or public surveys that use multiple dimensions and rely on preferences that are not necessarily easily interpreted as dollars.

The Benefits Approach

While money is often the obvious common metric. Another simple metric common to psychological measurement is that of how relatively satisfied people feel with alternative ways of allocating water and the alternative benefits that this allocation implies. For this reason a Water Benefits Accounting and Assessment (WBAA) methodology has been developed (Syme and Nancarrow, 2008) that attempts to establish and apply such a metric. In short, it assesses individuals' attitudes towards the importance of a variety of benefits the community feels it can gain from a particular body of water and the degree to which people feel they are achieving those benefits under current conditions. People are then asked the same questions in regard to what they feel would happen to the achievement of each benefit if a particular policy were adopted. In this way, the change in score will measure the overall potential increase in benefits if alternative policies or plans were adopted (see McIntyre et al., 2006, p22). Ideally also the 'do nothing' case should be investigated to accurately assess the value of any policy. That is, if no plan or policy was introduced, what would be the effects on the delivery of benefits over time due to current pressures on the water resource. Is there a stable condition or are things likely to improve or deteriorate if nothing is done?

More particularly, in the case of an adopted plan and limited resources to implement it, a benefits map can be developed which will show the decision makers and the community which priority actions will maximise the benefits outcome from the implementation of the plan. This benefits plan can provide very useful input into the debate about priorities for implementation. This debate will also, of course, have to consider issues such as the resources, skills, and state of knowledge available to implement particular activities. This methodology is easily adaptive to catchment wide issues where recreational experience provide a considerable number of the possible benefits. This was demonstrated in a study examining the benefits associated with the implementation of the Lake Mulwala Land and On-Water Management Plan on the border of Victoria and New South Wales. Prior to the implementation of the plan it had been the subject of severe community controversy and the benefits map provided an agreed template upon which to move forward.

Another advantage of taking this approach is that because interest groups and demographic and locality information is collected at the time of the interviews to identify the benefits, WBAA is capable of demonstrating which parts of the community will perceive different benefits from water plans and for what reasons. It will also do the opposite in being able to define negative impacts on some sectors of the delivery of benefits to particular groups. Potential conflicts about fairness and equity issues can consequently be collected in an open and quantitative way. Benefits analysis provides a visible representation of all values and viewpoints during decision making.

Environmental criteria that underpin water benefits are also identified to target specific issues such as sedimentation, erosion or maintenance of landscape features such as forest cover or existence of native flora and fauna (Moran et al., 2004).

Multi-criteria Analysis, Multi-Attribute Utility Theory and Analytical Hierarchy Processes

All of these processes preferably in a consultative framework identify the significant criteria that will be the determinants of the most preferred alternative. Multi-criteria analysis MCA can be a rapid process whereby the key criteria for policy or management evaluation can be selected and altered to arrive at a most preferred solution for recreation and environmental issues (Proctor and Drechsler, 2006). Multi Attribute Utility (MAU) and Analytical Hierarchy Process (AHP) methodologies tend to nest specific attributes or criteria under general headings and their importance to an overall judgement of an existing or a new policy rated under these headings (Saaty, 1999; Weiss and Weiss, 2009). Thus for the analysis of particular policies, specific attributes and criteria can be altered to arrive at the best possible outcome both for the individual and the group. In the case of individual differences, group negotiations can be undertaken to arrive at an agreed best position (Hajkowicz and Collins, 2007).

The advantages of MCA are that agreements can be reached quickly and simply on the key issues whereas the MAU and AHP processes can become more time consuming as each criteria and attribute is more closely defined. Nevertheless, all can be used to assist expert judgment in developing catchment indicators where causal relationships between different factors can be included along with importance ratings (Randhir et al., 2001; Randhir and Shriver, 2009) and more values-based weightings from stakeholder and community involvement. It must be noted that these processes do not necessarily arrive at the most economically efficient solution, but all can provide a consensual basis for planning and can be used to ensure a degree of procedural justice in the planning process itself (Syme and Nancarrow, 2005). Procedural justice (or a perception that the decision making process is fair) has an often demonstrated correlation with the acceptance of the decision (Lawrence,et al., 1997, Tyler and Blader, 2000).

These types of approaches are frequently applied to risk related environmental policy generally (Ananda and Herath, 2005; Schultz, 2001) and can be adapted somewhat to provide consistent data with the CRCWQT work using the water resource risk assessment framework (Miller et al., 2006). They are also useful in establishing management priorities for catchment management for watershed protection. All have been used for risk based decisions both in the public and utility risk judgement area (e.g. Edwards and Winterfeldt, 1987). They are recommended practice for strategic planning for utility risk management (Pollard et al., 2004; Rothstein and Kiyosaki, 2003). These techniques have potential application to the catchment drinking water catchment issue.

2.1.6 Relating Policy Assessment to Ecosystem Services.

In the catchment protection literature there is significant work that suggests that before catchments can be used for other than drinking water purposes, the undisturbed catchments' contribution to ecosystems services should be considered. There is a significant literature on the need to protect catchments not only for drinking water supply, but also their ecosystem services, protection of biodiversity and landscape amenity (e.g. Leder et al., 2002; Postel, 2005).

Ecosystem services are provided by both 'natural' and modified ecosystems and only some are incorporated into markets. They include:

- provision of food, fresh water, wood, fibre and fuel;
- nutrient cycling, pollination and soil formation which support primary production;
- regulation of climate, floods, disease and water purity;
- aesthetic, spiritual, educational, recreational and cultural aspects.

The report by 1400 leading scientists of the Millennium Assessment (MA) (2005) highlighted the global decline in ecosystem services while expanding populations are placing new demands on them.

Many ecosystem services are neither priced nor valued in the economy and they have been exploited and ignored. We urgently need to bring these services into the mainstream economy. There is now a national program to link ecosystem services through the Australia 21 Activity. Australia 21 (2008) seeks to bridge the gap between research and policy to create new 'frameworks of understanding' that can be applied to the development of public, corporate and community policy outcomes. With the support of three State Governments, Australia 21 held a series of expert roundtables and developed a national ecosystem services strategy. The aim of the strategy is to weave ecosystem services into our social, cultural and economic life by 2020.

According to the Millennium Ecosystem Assessment (Millennium Assessment, 2005), ecosystem services are:

> *"the benefits people obtain from ecosystems"*.

These benefits include vital services such as water flows, nutrient cycling and biomass production that underpin all our production systems and lifestyles. The Millennium Ecosystem Assessment, identified four categories - provisioning, regulating, cultural and supporting ecosystem services and linked them to human well-being (see Figure 2).

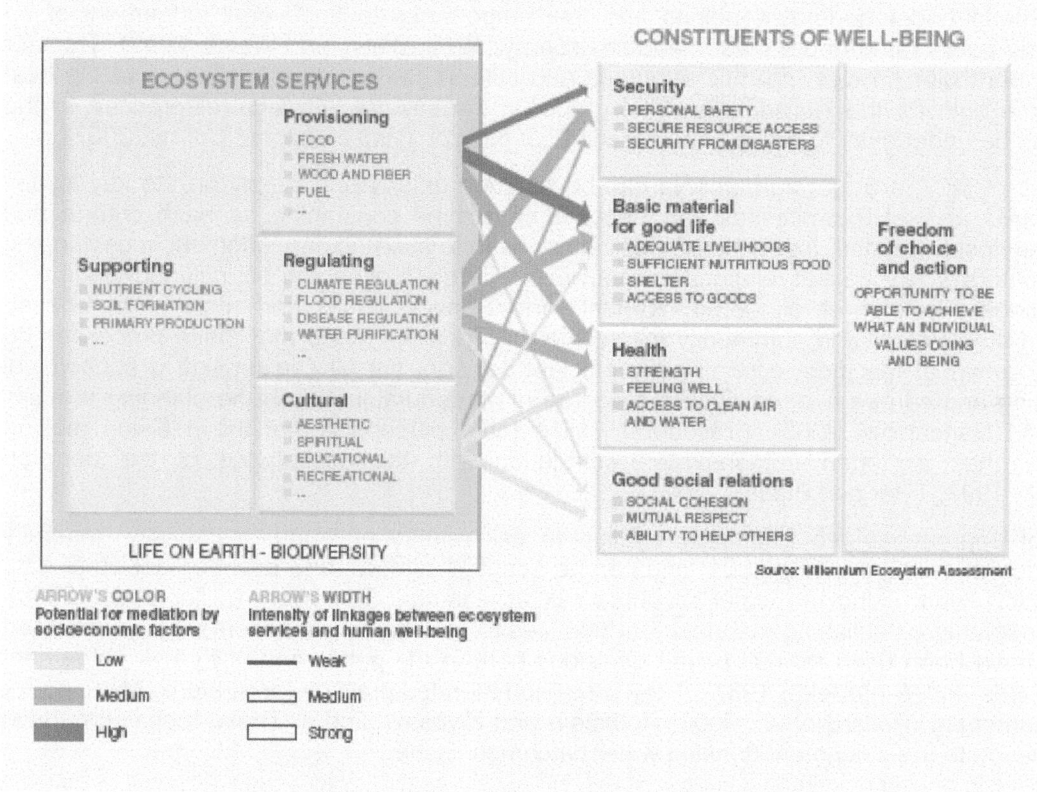

Figure 2. Links between ecosystem services and constituents of well-being (MA, 2005)

The above definition is supported by Australia 21. While other definitions exist, the simple definition for ecosystem services above provides a broad scope for interpretation by local communities to devise measurement systems (metrics) that are meaningful to them and best suit their needs. The figure categorises recreation as a cultural ecosystem service along with the provision of fresh water.

While there is a clear push to encourage as many market mechanisms to create an economic value for services, it is evident that there are a number of challenges to relying entirely on market mechanisms at this stage. These include difficulties with arriving at a classification systems to deal with all natural resource management problems (Fisher et al., 2009), problems of dealing with utilitarian versus non utilitarian services (Egoh, et al., (2007), pragmatic issues of being able to identify when government intervention rather than markets are appropriate (Kroeger and Casey, 2007) and limitations in our current ability to provide an inclusive definition of the links between ecosystem and human health (Castle, 2000).

While many authors emphasise the "ecological services" provided by protected environments, it is also important to consider the indigenous cultural values provided by catchment protection. Burger et al. (2008) suggest that such values should be included as a basic component of our definition of natural resources. These authors suggest that traditional indigenous pursuits such as hunting and camping are enhanced by protected environments, as are highly valued or sacred sites. It is likely that active and passive indigenous use of land will correspond to access to water. While a drinking water dam in itself may be regarded as a disruption of such activities in some circumstances, the RPZ and outer zone protection has the opportunity to retain many aspects of the wider definition of natural resources offered by Burger, et al, (2008).

Thus many of the issues surrounding non-market valuation are embedded in wider definitional, implementation and institutional issues. Nevertheless, in developing an ecosystem services framework for the catchment protection issues, links can be drawn between recreation, drinking water protection and human wellbeing. Such a systematic approach can provide the context within which effective social research programs can be conducted.

2.2 Conclusions

The Logue Brook controversy and the subsequent stakeholder and social research have demonstrated that, to serious recreators, the loss of a recreational facility because of drinking water protection is of considerable concern. The general public however, when asked about the same time gave great priority to the provision of pure drinking water. This latter finding is not surprising and has been shown consistently to be a top priority in many surveys over the years. In this review we have presented some psychologically based reasons as to why these differences may be the case. Serious recreation leads to site identity, difficulty in providing substitutes and a feeling of endowment. This does not necessarily mean that the issue can be simplified to one of gradually rising recreational demand and inevitable conflict.

In this review we have presented a series of socially based research approaches that can be linked to provide models of attitude which can enhance the understanding of values, risk assessment, the dimensions underlying acceptance of catchment protection policies, and the wider systems frameworks within which these data can be interpreted. We also show that both economic valuation and less utilitarian approaches are required to assist policy development. It is also stated that these techniques need to be interpreted within a process which involves stakeholders providing their priorities, and understanding risk and uncertainty (especially in the light of climate change) as well as dealing with the aspirations of the public as a whole.

Not all of these methods can or will be used within the current project, but by creating a coherent view of how the research fits into an agreed system or ecosystems framework, we can both efficiently and effectively conduct a social research program which will facilitate current decision making on water resource protection but also provide direction for the future.

3. SCOPING PHASE

The purpose of the Scoping Phase is to gain *an understanding of the key issues* associated with the local context of drinking water catchment protection policy. This is done in a number of ways:

- understanding the issues from the perspectives of the *range of stakeholders and professionals;*
- understanding the relevant legislative framework that underpins drinking water source protection and recreational access to drinking water catchments; and
- determining cost implications of any changes to source protection and recreational access to drinking water catchments associated with the requirement for different levels of additional treatment.

The major feature associated with understanding the different professionals' and stakeholders' perspectives is *listening*. This is not an attempt to find representative viewpoints, or to educate, but to elicit the range of issues and perceptions associated with the current and any proposed new policy so that there are "no surprises" at any stage of the project. This can be done by individual interviews and small focus groups. The mix of these techniques with the different groups may vary, depending on the local context. The Scoping Phase will help develop the context specific questions in the Survey Phase of the project so that meaningful data can be collected.

In the WA context, this phase asked professionals and stakeholders their thoughts and attitudes about the current policy of catchment protection and their thoughts about any relaxation of recreational access to the drinking water catchments. This was done by individual and small group interviews (2 to 4 people) with professionals and focus groups with the general public and members of formal recreation groups. These two techniques are described below. Overviews of the findings are provided in this section with full details in the Appendices.

3.1 Interviews with Professionals

Interviews with professionals should be free ranging and elicit information according to the particular relevant expertise. Relevant professionals include, but are not confined to, people in the areas of drinking water production and regulation, catchment management, management of parks and forests, tourism, recreation, planning, environmental science, ecology and so on. Eight to fifteen interviews would be sufficient, depending on the local issues.

Interviews can start with an explanation of the project, and then ask a general question such as, "What do you think about the current policy of ...?" A *conversation* will ensue with questions asked to draw out pertinent points as they arise rather than through a prescribed set of questions. The interviewer can have a "checklist" of points that need to be covered and refer to this periodically to ensure this occurs. Such a checklist could include:

- Treatment and health
    - Minimum requirements
    - Dealing with water quality incidents
    - Challenges with changed recreational access to catchments
    - Health benefits of recreation
- Risk
    - Known and unknown risks
    - Management or prevention
    - Relationship to recreational access

- Governance and Management
    - Different responsibilities and workability
    - Relationship between agencies
    - Policing
    - Equity and cost issues
- Planning
    - Synergies with and impediments to future developments (e.g. urban and tourism)
    - Justification for current policy
    - Alternatives for recreation
- Recreation
    - Organised groups v individuals
    - Relationship between organised groups and governing agencies
    - Relationship between groups and activities
    - Substitutability
- Ecology
    - Opportunities for protection of biodiversity
    - Preservation of endangered species
    - Relationship with recreation
        - potential for wilderness experience
- Future Issues
    - Population growth
        - Managing future recreation demand
    - Climate Change
        - Impact on drinking water sources
        - Demand for water-based recreation
        - Demand for agricultural/horticultural irrigation
- Any other issues

3.1.1 Overview of Professionals' Interviews in the WA Case Study

The following is a brief summary of the major outcomes of the interviews with professionals in the WA Case Study. Detailed notes can be found in Appendix 1.

It was said that the last health scare related to drinking water quality had occurred in Perth over 100 years ago, and as a result, the community had little knowledge of water quality incidents and was confident about the safety of the water that came from their taps. The current policy of catchment protection allowed for minimum treatment of the water for drinking, and that there was no question that having to rely on advanced treatment for safe drinking water was a "second best option".

The number of drinking water dams allowed individual dams to be taken off-line if there was any suspicion of a water quality issue or incident. It was said that it was preferable to minimise risk rather than to mitigate risk and that the Australian Drinking Water Quality Guidelines recommended *preventative* risk management. There was a lack of understanding about micro-biological contamination and that the drinking water catchments could not cope with body contact with the water. There was concern that the authorities had no knowledge of how people would react to a drinking water incident and that it would take a long time to regain public confidence.

There seemed to be no understanding of or justification for the choice of 2kms for the RPZ, but that it seemed to "have worked". However, it was difficult to argue the reasoning for the 2kms with the recreators.

There needed to be better planning for recreation overall and perhaps classify catchments according to vulnerability. It was thought that recreational facilities at other water bodies should be better developed and then decide what recreation could occur inside the drinking water catchments, and what activities were compatible.

There was considerable discussion about the workability of the policy and its By-Laws and inter-agency cooperation and coordination. It was said that there was a need for more consistent Regulations with greater ability to deter wrong doing, and a need for a one-stop-shop for recreational permits. There was also a lot of discussion about the difficulties of effective policing of the RPZ and the outer catchment. It was thought that each dam should have a management plan developed with the participation of all relevant Government Departments and the community.

It was said that organised recreation groups generally followed the rules, but the problem was with individuals and the "hoons".

There was support for the RPZ as habitat for endangered species, and that there is ecological and social value in having undisturbed areas.

It was suggested that with future climate change, the nature of the dams may change with less emphasis on some as sources of drinking water. There will be greater emphasis on manufactured water (desalinated and recycled).

Overall, it was asked "why go backwards from a proven system?" even though it was acknowledged that there was not enough data to answer some of the questions yet.

3.2 Stakeholder Focus groups

3.2.1 General Community Discussions

The purpose of this qualitative data collection exercise is to gain an understanding of the ***range of views*** of the "general community" in relation to drinking water protection and recreation in drinking water catchments. With the intention of gaining the participation of between 10 and 12 community members, it is suggested that about 16 potential participants be recruited. This can occur through random telephone calls and explaining what the project is about and what is expected of the discussion. It is best that this recruitment be conducted by an impartial party, as also should occur for the conduct of the focus groups (i.e. not the water utility nor any of the government agencies associated with the policy or issue). When recruiting the community members, discussion should occur with the consultant as to the use of any maintained community panels for recruitment of participants in such events. If a panel is used, it should not be "water interested" people, nor anyone who has participated in a focus group over the previous six months.

The sample of community members can be stratified to obtain a mix of age groups, genders, socio-economic status and geographic locations. It can also be required that certain professional expertise not be represented, such as health experts or environmental scientists. In terms of geographic locations, it would be useful to have people who lived near and far from the drinking water catchments, and people who have other recreational choices, such as close to other water bodies or the coast. Participants should be offered a small incentive in recognition of their time, such as a \$60-\$80 gift voucher or cash, and perhaps assistance with the cost of parking if applicable.

Potential participants should be advised that they need no special knowledge as we are interested in what everyone thinks. This attitude should be maintained throughout the focus group and only basic and factual information provided with no attempt to persuade or educate to any particular viewpoint. It is important to know what people think and how they talk about the issues.

Focus Groups should be held on a Monday to Friday evening or on a Saturday morning or afternoon to maximise the range of people who can attend. Apart from the facilitators, a couple of utility or agency people should be present to provide some basic background information and answer any technical questions that might be asked.

Discussions should occur in two small groups of between 5 and 8 people in each, depending on the numbers who actually attend on the occasion. Participants can be pre-assigned to the groups to ensure a mix of genders, ages and locations at each. Each group should have a facilitator who takes the groups through a series of pre-determined questions. Digital recording of discussions is preferred if participants do not object.

The following is a suggested format for an evening focus group, utilising the questions asked in the WA case study. These questions may change according to the local context. While allowing the discussion to be comfortable and encouraging of people to speak, strictly adhering to the order of questions may be difficult. However, all points should be covered by the end of the evening.

6.00pm Introduction and Explanations (10 minutes)

Facilitators introduce themselves and their roles and also introduce the utility or agency officers. It should be advised that the officers are there to provide information and answer questions about drinking water catchment areas and treatment of water for drinking.

Facilitators then provide a brief overview of the issue and the project and what the expected outcome of the discussions will be and how this will contribute to the overall project.

6.10 Presentation of Background Information (10 minutes)

As noted previously, this should be short and provide sufficient information for people to be able to contribute to discussions without trying to influence them or educate to any particular position. The presentation should be clearly factual, and where possible provide other examples or authorities outside the local context. Suggested points to cover include:

- What is a catchment and what is the current policy of catchment protection?
- How is water currently treated and delivered from the dams to households?
- How are catchments managed and water treated in other places nationally?
- What are the health and environmental implications of opening catchments to recreation?
- What are the treatment implications of opening catchments to recreation?

6.20 Questions (10 minutes)

6.30 Group Discussions (30 minutes)

Q1. *Have you been aware of this policy previously? – What do you think are the good aspects ..... and the bad aspects?*

Q2. *If you were in charge of the policy would you leave it as it is, or would you change it? Why? ..... How?*

Q3. *What do you do for recreation? What sorts of things do you do? Where do you do it? Would you benefit from the catchments being opened up to more human activity?*

7.00 Break & Refreshments (20 minutes)

7.20 Group Discussions (50 minutes)

Q4. *Some people totally agree with the policy and believe catchments should be kept pristine for total safety of drinking water.*

Some people believe that a few more activities should be allowed closer to the water (e.g. bushwalking or horse riding), but not in the water ----- or perhaps allow no more contact with the water than fishing from the edges.

Some people want the catchments totally opened up for all activities and have greater treatment to ensure safe drinking water.

Let's talk around those three points of view. What are the advantages and disadvantages of each? Which point of view do you mostly agree with?

Q5. *There are some who say that those who want to recreate in the Reservoir Protection Zone can find other places instead.*

Others say that with increasing population it's harder to find places to go that are close to where they live, and recreation is important for good health.

Let's talk about this. Are there other places that people can go? What are the health trade-offs – healthy lifestyle/healthy drinking water?

Q6. *It is said by some that there are costs to **not** using the land - development and tourism loss - fewer recreational opportunities and so increased health costs if it is harder to find other places to recreate*

It is also said by some that there are costs to using the land - loss of pristine land: threats to biodiversity; loss of protected habitat for endangered species; environmental damage such as erosion - increased fire risk and hence potential to contaminate the water - increased financial costs: treatment infrastructure; more sophisticated treatment; greater energy costs associated with the greater treatment.

What is important to you if you had to make the decision about using or not using the land?

If the land was opened up, who should pay for the management and treatment?

8.20 Close (10 minutes)

Thank the participants and advise what happens next and what will happen to their input. Distribute incentives and obtain participant signature that it has been received.

3.2.1.1 Major Points from the WA Case study

The following are the key points that emerged from the WA case study discussions with the details provided in Appendix 2.

All of the participants were recreators of some kind with some being members of formal recreation groups. Most had heard of the policy previously and it was thought to be "common sense".

> *"We pride ourselves on safe drinking water, so we shouldn't jeopardise it."*

The minimum treatment that was required for their safe drinking water was appreciated by the groups, and it was the avoidance of added chemicals that was important to them, more so than the additional cost of advanced treatment.

> *"We've ruined our food with chemicals, we need to keep the purity of our water. We don't want added chemicals and added treatment."*

There was a consistent message that there should be no compromise on the policy that would jeopardise the safety of the drinking water they all enjoyed. It was noted that it was a public good, and that minority groups should not have too much say. While some thought that perhaps some additional activities might be allowed in the 2km RPZ, it was felt that the longer people could be held off from doing so, the better it would be. It was suggested to use the extra money saved from not needing additional treatment to employ additional rangers.

While it was thought that people could drive an extra 100kms to recreate at other places, it was also agreed that there needed to be better planning for recreation. This would also benefit tourism. It was said that people were unaware of alternative places to go and this should be better communicated. While people were wanting "wilderness experiences", opening up the catchments would not provide this as there would be more people on the catchments. There was also acknowledgement that perhaps they were looking at it only from a city perspective, and that a local body of water for recreation might be important for rural people.

Overall, it was agreed that purity was most important and that the areas around the dams were minimal areas to give up.

> *"What are we willing to pay: financial, environmental, health. We always wait until things are critical before doing something about it. Prevention is better."*

3.2.2 Recreation Group Discussion

A similar process to that of the community should be conducted to obtain the range of views held by members of formal recreation groups. While it is not intended to provide an opportunity for all groups to participate, it is important to obtain a mix of land and water, passive and active recreation clubs or groups such as some of the following. In this regard, it is the mix of activities that agree to participate that is important rather than actual numbers.

Passive Land:

- Bushwalking
- Rogaining
- Orienteering
- Bird watching

Active Land:

- Trail biking
- Mountain biking
- 4 wheel driving
- Hunting/shooting

Passive Water:

- Freshwater fishing
- Canoeing/kayaking

Active Water:

- Water Skiing
- Motor boating

Particular recreation groups and contact people may be known to the utility or relevant agencies, but details can easily be found on the web. Specific groups should be contacted by email or mail, and invited to provide one participant for the discussion. Numbers should be kept to a maximum of 10 and all participants should have the opportunity to hear the perspectives of others, so discussions are best to occur in a round table format.

While the facilitators should attempt to follow the program as outlined for the general community, they should be aware that these discussions are likely to be more "spirited" and knowledgeable than occurred with the community. Participants will bring with them a history of previous interactions with Government and they will probably have pre-determined points they will want to make. It is important that they be able to do this, so the questions will probably provide more of a checklist than a strict order of proceedings.

3.2.2.1 Major Points from the WA Case study

The following are the key points that emerged from the WA case study discussions with the details provided in Appendix 3.

Each of the participants provided a brief overview of their group's perspectives at the start of proceedings. The bushwalkers, freshwater fishers, and orienteerers all outlined previous agreements with Government that had allowed their access to the RPZ, all of which had been repealed even though they were of the understanding that Policy 13 would honour all prior agreements. This was a major point of contention.

There was considerable discussion on policing the policy with the suggestion that more people in the area would mean that some would be less likely to pollute and desecrate. The notion of self-policing was thought to be both effective and non-effective. It was said that the Rangers only worked from Monday to Friday and that it was the weekends when they were needed.

The problem of managing future demand with the limited amount of prime quality land for recreation and increasing population was highlighted. Current recreational areas were said to be already "full" (e.g. walking areas in Perth) and it was questioned as to whether more areas would draw from demand or add to demand. This was thought to be both a spatial and temporal problem. There needed to be consideration of what activities could co-exist, and when and where others could go. It was said that a Master Plan was required with a multi-objective approach to maximise the utility of the area, rather than maximising the water quality. The Plan needed to address how best to utilise the land for recreation while protecting the water. In managing demand it was said that all the groups needed to work together and with the government departments.

There was considerable discussion about the difficulties of interacting with the different government departments with experiences of conflicting advice and requirements. They called for one coordinated body to deal with. It was thought that there was a lot of misinformation about recreation and its risk to water quality and this was not so if managed properly. There were international models for management and education that could assist.

It was stated that there were other allowable land uses (e.g. mining) that were more destructive of the land than was recreation. Landscape was important to recreators, so it was preferable to allow recreation to other developments. The economic benefit of recreation was ongoing, where for such as mining, it ended when the mine ended.

In discussions of costs and who should pay, it was agreed that there were social benefits from recreation that flowed on to the community, such as a socially stable community, and a healthy community. Therefore, society could pay.

3.3 The Relationship between Sport and Recreation Policies and Drinking Water Source Protection

Naturally this will vary from context to context, but it important in providing a thorough understanding of the issues associated with the policy or potential policy changes. Following, as an example, is the outcome from the WA case study.

Hughes et al. (2008) provides a formal review of the relevant legislation and Departmental roles in drinking water catchment management and protection. There appear to be significant overlaps in responsibility and at times confused relationship between agencies (e.g. the CALM[1] Act, 1984 with management powers for the Department of Environment and Conservation (DEC); Country Areas Water Supply Act 1947 – CAWS Act which provides "control of catchment areas" but no reference to management function and no provisions on how it relates to the CALM Act). The State-wide Policy No. 13[2], issued in 2003 by the Department of Water (DoW), "was not created on the authority of any legal powers granted to the Commission [3] by state legislation or otherwise, it is not legally binding on other governmental agencies and departments, but may be used as a guideline. Policy 13 states that all activities proposed on Crown land require approval from the land manager, which in most instances will be DEC." (Hughes et al., 2008).

In pragmatic terms the Department of Sport and Recreation (DSR) has no formal control over catchment land use or the human activities allowed on them. Much like the recreators themselves they operate within the constraints and guidelines of Policy 13. While recreation is forbidden within the RPZ apart from historical uses which do not affect drinking water quality, there are a variety of largely passive activities that can be conducted in the wider existing and proposed drinking water catchments. These are conditional on approval from the DEC, the DoW and the Water Corporation. Often it is the experience of recreators, and presumably the DSR, that conflicting rulings can be made from time to time between the differing agencies when permission for recreation activities is requested.

By and large, the DSR acts as an Advocate for recreation interest but acknowledges that "new models for industry representation and consultation are required to achieve effective strategic leadership and effective policy influence" (DSR, 2006). This may be a challenge for all agencies if adequate planning for water protection and recreational provision is to be achieved. Currently there is no formal plan for the key catchments where water protection and recreation interact.

DSR recently commissioned a review of drinking water catchment protection and recreation and have provided a submission to the current WA Government Inquiry into protection of drinking water catchments. This would seem to indicate that the issue is of mounting concern for recreational planning. This is especially the case in view of the drying climate in the South West of Australia and the increasing demand for water and nature based recreation close to the city.

The findings of the DSR review were presented to the Parliamentary Inquiry (http://www.dsr.wa.gov.au/assets/files/Presentations/Recreation%20exists%20and%20can%20be%20managed.ppt) and was entitled "Recreation Exists and can be Managed".

The presentation provided a "Wish List" which provided some surprises in view of previous stances.

- No water-based recreation in drinking water reservoirs
- No land-based recreation in vulnerable catchment areas.
- No roll-back of existing recreation access for irrigation dams.
- A funded process to plan and implement a zoned system of nature-based recreation opportunities including:
    - Commitment, legitimacy
    - Dialogue and process

---

[1] Conservation and Land Management

[2] Water and Rivers Commission, 2003, Policy and Guidelines for Recreation within Public Drinking Water Source Areas on Crown Land

[3] The previous Water and Rivers Commission

- Development of a strategic framework
- A risk-based approach to identifying spatial and temporal zonings
- Icon 5-star facilities
- Excellence in management
- An effective funding model
- An effective governance model that includes public, private and commercial partnerships.

Among the conclusions were the points:

- DSR in supporting the current policy of exclusion of water-based recreation in drinking water reservoirs is helping the Water Corporation save $millions in treatment infrastructure and operating costs.
- It is also helping to significantly reduce its liability exposure.
- As a trade off the Water Corporation may consider providing resources for planning and implementing alternative nature-based recreation opportunities.

  While there is a strong interest in drinking water protection by many, there are a number of unknowns that deserve further attention if mutually agreeable solutions are to be reached with agencies that differ in their approaches to risk and uncertainty. The water based agencies currently operate on risk avoidance whereas others adopt a risk management approach. There is a need for holistic long term recreational planning for the catchments not only in terms of recreation but also acknowledging competing land uses. This will require greater resources than are currently available.

Information required to provide a more evidence based planning and negotiation include quantification of:

- Current and potential demand
- Potential demand if restrictions are relaxed
- Non formal recreation
- Historical recreation patterns
- Illegal recreation in the RPZ
- Recreation activities where substitutes can be identified
- Current uncertainties for protecting drinking water quality in the light of a drying climate and non rain dependent water sources
- Potential cultural and indigenous benefits of Drinking Water Catchment Protection
- Ecological and Biodiversity benefits of catchment protection
- The health and economic benefits of recreation

3.4 Cost Implications for Additional Treatment Associated with Recreation in Drinking Water Catchments

This task will again be dependent on the particular context and the policy development/changes being considered. Any changes to recreational access to the drinking water catchments would supposedly result in changes to required treatment and hence changes to associated costs. To allow realistic consideration in the Survey Phase of the implications of any policy changes, these costs need to ultimately be presented as cost increases to households. This is a task for the utility and it needs to be done in a timely fashion for easy incorporation in the Survey. Following are the total cost implications determined for the WA Case Study. The costs were then expressed as additional percentage cost increase for households in the survey, and actual dollar increases for a household of average consumption were also provided as additional information if required.

3.4.1 The WA Case Study

The Legislative Council's Public Administration Standing Committee is reviewing Policy 13; Recreation in Drinking Water Catchments. As a part of the Water Corporation's submission, the approximate cost implications of allowing recreation on drinking water sources have been estimated. Indicative Metropolitan & Country examples are provided here.

3.4.1.1 Issues

When used for drinking water, both the dam and the catchment play important roles in the management of drinking water quality supplied by the Corporation.

The existing source protection approach is to largely avoid risk through catchment protection zones; this will be reduced in effectiveness by additional catchment based recreation as new risks associated with this recreation will be introduced.

The effectiveness of the storage barrier will be lost through water based recreational activities in feeder streams or on the reservoir.

Therefore recreation on either the catchment or the water body will compromise the Corporation's management of drinking water quality at the source.

The Corporation advocates that changes to the public health risks associated with drinking water should be minimised in response to any proposed increase in recreation activity in catchments. Additional water treatment does offer a way to manage additional risk.

However, an environmental barrier (preferably storage) after treatment is required if the public health risk change is to be minimised.

The treatment requirements for water subsequently drawn from the secondary storage must also be reassessed.

It will be necessary to continue to manage some existing sources with a high level of protection in order to offer an uncompromised storage barrier for receiving treated water from other sources where catchment risks are permitted to increase.

Therefore transport and storage to a suitable location must be both possible and included when considering the cost implications of allowing recreation in the source.

Issues associated with ongoing increased management costs for recreation catchments retained by the Corporation, assignment of responsibility for asset ownership for recreational facilities and any residual asset disposal issues (dam safety) are not included in this assessment.

3.4.1.2 Metropolitan Example – Multiple Sources

North Dandalup, Conjurunup, Lower South Dandalup & South Dandalup Dams are Metropolitan Dams in close proximity to each other and could be considered for central treatment (Microfiltration & Ultra-Violet Disinfection) and bulk water transfer to Serpentine Dam. Additional outlet and treatment capacity from Serpentine to replace the Dandalup trunk main supply capacity is also required.

Costing of this scheme (Table 1) allows comparison with a source abandonment and replacement strategy for this significant group of sources. Only capital and operating costs associated with the new assets required to implement the treatment-to-mitigate-the-risk option are captured at this time.

Table 1. Costs – Metropolitan Example

Accommodating Recreation on Dandalup Sources: Costs		**Abandonment NPV Cost**
Capital	$346 M	**$545 M***
O&M (NPV)	$203 M*	
TOTAL	**$549 M**	

* NPV Costing assumes that the current dry climatic conditions continue. The NPV would increase were wetter conditions to be experienced as more water would be treated by the Water Treatment plants.

3.4.1.3 Regional Example: Single Source with no Environmental Storage Barrier

Harding Dam is one source for the West Pilbara Water Supply System; this connects Karratha, Roebourne, and a number of other small towns. The Water Corporation does not support any high risk recreational activities in drinking water supplies that does not include a natural barrier. *No such storage barrier is included in this costing; therefore these costs must be considered in light of a failure in treatment placing the entire water supply at risk.* The possibility exists to use Millstream aquifer as this storage barrier. This technically complex option is currently being costed.

Water treatment at Harding Dam is based around the process of Microfiltration; it includes a series of process steps to ensure the disinfection and removal of natural organic matter to minimise the creation of disinfection by-products (THM's).

Additional treatment barriers (Dissolved Air Flotation, Biological Activated Carbon & Ultra-Violet Disinfection) would be required to deal with the increased pathogen and water quality challenges posed by recreation on the Harding Dam. The cost of the required capital upgrades and the ongoing increased operational costs of the treatment plant are included in Table 2; *they do not factor in the requirement for transfer to a storage barrier*.

Table 2. Costs – Country Example

Accommodating Recreation on Harding Dam: Costs	
Capital	**$50 M***
Operating (Yearly)	**$3 M***
TOTAL (20 Years)	**$110 M**

*Does not include existing capital ($40 M) and operational ($2.7 M) costs associated with the existing complex treatment.

4 SURVEY PHASE

The aim of the Survey Phase of the project is to investigate: knowledge of drinking water sources and acceptance of the current policy of catchment protection; risk tolerance; attitudes to uncertainty and potable water purity; aesthetic and landscape preferences; trust in water treatment, managers and science; the relationship between location and substitutability of other places for the drinking water catchments for recreation; lifestyle and recreational preferences; acceptability of alternative policy options and attitudes to cost and who should pay and demographic data that has been shown to influence attitudes to risk.

This phase of the project involves similar surveys of three different samples and builds on the findings of the Literature Review and the Scoping Phase.

- General community
- Members of formal recreation groups (referred to below as recreators)
- Professionals associated with drinking water, catchment management and recreation

The WA case study used two different forms of delivery of the questionnaire for practical reasons, but this had little effect on the outcomes. The questionnaire was delivered to the community and the recreators through personal telephone interviews of approximately 20 to 30 minutes in duration, and to the professionals through a web-based survey. The web-based format required a little adjustment of the questionnaire from the first, and this should be acknowledged in the analysis. However, it was considered that professionals would be more likely to participate if they could do the survey in their own time, and have the ability to do so in several sittings if required. A web based format for the other samples is not recommended.

Both processes should be conducted by an independent data collection agency so that respondents can be assured of their anonymity, especially as the researchers will be familiar with some of the recreator sample and also with those in the professional sample. However, the researchers should ensure the agency is thoroughly briefed on the questionnaire, and should attend the interviewer training session.

Due to the length of the telephone interview, community and recreator respondents should be offered an incentive in recognition of their time. In the WA Case Study, this entailed entry into two draws, each for $300 of department store gift vouchers[4].

Details of the recruitment of the three samples follow. Details of example samples from the WA Case Study can be found in Appendix 4.

4.1 The Three Samples

4.1.1 Community Sample

Assuming a metropolitan area is involved, a stratified sample of 600 random community whose drinking water is supplied from the catchments under question should be targeted. This should be achieved by equally assigning respondents to geographical locations defined by relative distance from the drinking water catchments. Areas that offer substitutability, such as coastal areas, should also be targeted with the assumption being that these people have more easily accessible alternatives to the drinking water catchments for their recreational pursuits. See Table A4.1 in Appendix 4 for the WA Case Study example of sample locations.

Half the sample should be male and half female, and half under 40 years of age and half over 40 years of age. All respondents should be over the age of 18 years. Community panels should not be used, but potential respondents drawn randomly from telephone lists or random number generation.

4.1.2 Recreator Sample

A sample of 100 recreators who are members of formal recreation groups should be aimed for, with the emphasis on obtaining a *range of active and passive, and land-based and water-based activities.*

[4] These were left over from the Scoping Phase as sufficient incentives had to be purchased in the unlikely case that all invitees would participate in the focus groups.

The recreation groups should be contacted and asked if their members would be willing to participate and if so, to provide the names and contact details of up to 10 members from each group. The goodwill and cooperation of the groups is therefore required and it may not be possible to obtain the target sample number as responding to the survey may not be a priority for the groups' attention.

However, the sample number should be sufficient to allow valid statistical comparisons while providing responses by the range of activities as noted above[5]. The survey data base should not identify respondents by recreation group to ensure confidentiality as numbers in the groups will be quite low.

4.2.3 Professional Sample

A range of professionals of relevant disciplines from the public and private sectors should be invited to participate in the survey through a web based format. The relevant disciplines will be associated with planning and management of drinking water dams and catchments such as, water managers, health professionals, catchment managers, sustainable tourism and recreation planners, academics and scientists.

A minimum target of 30 professionals should be aimed for, but this may not be achieved in reality as the pool of "experts" in this area is not large and attempts should be made to achieve a balance of numbers across the disciplines. A networking process should be used to identify people working in the area and they can be invited to participate by email and provided with background information on the project.

4.2 The Questionnaire

The following questionnaire consists of measurements (e.g. attitudinal measures) that are common to all case studies and sections that will be developed as part of the specific context. In these cases, examples from the WA Case Study are shown. The questionnaire is presented here in two ways:

- An overview of the measurement themes (Section 4.2.1)
- The structured form for delivery with instructions (Section 4.2.2)

Be aware that it is always tempting to want to put *more* rather than *less* in a questionnaire and to provide the respondent with "plenty of information" to ensure "informed responses". The purpose of a survey is to find out what people currently think about an issue and as such, they need to be representative of others in the area. Therefore, information provided should be the minimum to allow responses to the questions. The information needs to be short and succinct as there is a limit to what people can "take in" over the phone. The length of this questionnaire is the maximum of what is possible while still obtaining reasonable quality data.

The refusal rate for the community sample will be high[6], but response rate should not be confused with response bias. Research has shown a generally poor correlation between survey non-response and response bias (Groves, 2006). That is, just because there is a low response rate, this does not necessarily equate to biased data.

4.2.1 Overview of the Questionnaire

The following is an overview of questions to be included in the questionnaire. It is provided in this way to show the themes that are covered by the questions, but the final questionnaire will not necessarily follow this order. For example, for ease of interviewing and respondent answering, many of the questions that entail statements answered on a 5 point agree/disagree Likert Scale will be asked together, and in varied orders.

INTRODUCTION

A brief introduction for interviewers to explain to respondents that we are interested in their thoughts on the protection of drinking water and allowing recreation in drinking water catchments.

---

[5] Greater than 60

[6] Refusal rates for surveys such as this are generally between 65% and 80%.

PARTICIPATION IN RECREATION AND OUTDOOR ACTIVITIES

A series of questions to ascertain the respondent's participation in outdoor recreation activities: what and where

SUBSTITUTION, SITE ATTACHMENT & DEMAND

Statements to be answered on a 5 point agree/disagree Likert scale that cover: few or many activities; few or many places visited or available; ability to easily participate; degree of enjoyment in the light of demand.

KNOWLEDGE

Questions about knowledge of where their drinking water comes from, how their drinking water is protected, and if they have ever knowingly visited a drinking water catchment and where.
Both *perceived knowledge* and *actual knowledge* to be obtained.

DEGREE OF AGREEMENT WITH THE CURRENT POLICY OF CATCHMENT PROTECTION

Short information (e.g. maximum of 250 words) to be read to the respondent outlining the current policy of catchment protection followed by a question on how much they agree or otherwise with the policy, rated on a five point Likert scale. This is followed by an open ended question to explain the reasoning behind the rating.

TRUST IN INFORMATION

Question on how much they trust the key agencies or groups to provide reliable information about catchment protection and drinking water quality, answered on a 5 point Likert Scale.

STRENGTH OF ATTITUDES

Statements answered on a 5 point agree/disagree Likert scale to cover the dimensions of: *centrality; interest; intensity; importance; certainty*

Catchment Protection

Statements answered on a 5 point agree/disagree Likert scale to cover issues of catchment protection and allowing recreation in drinking water catchments.

TRUST IN AUTHORITIES

Statements answered on a 5 point agree/disagree Likert scale to cover issues of trust to provide safe drinking water; manage the provision of drinking water responsibly; keep abreast of water treatment science; and perceptions of the water utility as a part of the local community.

RISK

Statements answered on a 5 point agree/disagree Likert scale to cover perceived risk of recreation around drinking water; technological fix; natural treatment; comparative land uses.
Three part question:

a. Scale of perceived likelihood that something could go wrong
b. Scale of perceived seriousness if something went wrong
c. Scale of perceived control of authorities to stop something going wrong

...... if recreation was allowed in the catchments and additional treatment works were installed.

VALUES AND ETHICS

Statements answered on a 5 point agree/disagree Likert scale to cover intergenerational equity; environmental values; considerations of the right to good quality water.

FAIR PROCEDURES

Statements answered on a 5 point agree/disagree Likert scale to cover fairness of decision making and the treatment of people in decision making processes.

LATTITUDE OF ACCEPTANCE with the current policy and policy alternatives

Respondent is provided with a range of policies from the current policy through alternatives with gradually increasing risk to their drinking water and the implications for their household in maintaining safe quality water (e.g. increased cost). They are asked which are *acceptable* and which are *unacceptable* to them. Then, they are asked which is *most acceptable* and which is *most unacceptable.*

WHO SHOULD PAY?

Questions about who should pay for any increased requirement for treatment, and also who should pay given possible health cost savings to society from recreation opportunities.

DEMOGRAPHICS AND PERSONAL INFORMATION

Membership of an organised recreation group
Preference for greater access to drinking water catchments than they have now
Age
Family unit
Country of birth
How long lived in the area
Gender
Sample location

4.2.2 The Questionnaire

The following provides details of the questions as well as instructions and briefing notes for interviewers. The data collection agency will use this to program their computer system for the survey. Where there are differences in procedures for the web based survey, these are noted.

Where it is stated that options not to be given to respondents, the options still appear in the questionnaire to assist the interviewer and to save later coding. However, there will still be a requirement for coding of the "other" options. Also, options can be added or removed depending on the specific context.

☞ ***Introduce the survey. Interviewer reads the data collection agency's standard introduction and includes "we are interested in your thoughts on protecting our drinking water and allowing recreation in drinking water catchments".***

1. Do you participate in any form of outdoor recreation (for example: walking, picnicking, cycling, fishing, sailing and so on)?

   ☐ YES ☐ NO - 2 ☞***skip to Question 3***

   If Yes, What recreational activities do you participate in?

☞ ***<u>Do not read out options below</u> but tick as many as apply in each category or complete "other" for the appropriate category. Prompt for detail if too vague.***
In the case of the web-based survey for professionals, provide options for selection.

<u>A. Land based passive – individual/group</u>

- ☐ walking;
- ☐ bushwalking;
- ☐ hiking;
- ☐ running;
- ☐ orienteering;
- ☐ rogaining;
- ☐ cycling;
- ☐ mountain biking;
- ☐ picnicking;
- ☐ camping;
- ☐ climbing;
- ☐ bird watching;
- ☐ rock/mountain climbing
- ☐ other, specify..........................

<u>B. Land based active – individual/group</u>

- ☐ trail biking;
- ☐ 4WD;
- ☐ hunting/shooting
- ☐ other, specify ....

C. Sport requiring specific facilities

- ☐ football/soccer/rugby;
- ☐ basketball;
- ☐ netball;
- ☐ hockey;
- ☐ tennis;
- ☐ cricket;
- ☐ swimming in a pool
- ☐ other, specify ....

☛ ***If only sporting activities** **and no other categories** **– skip to Question 3***

D. Water based passive in the natural environment – individual/group

- ☐ swimming;
- ☐ fishing;
- ☐ canoeing/kayaking/rowing;
- ☐ sailing (no motor);
- ☐ wind/kite surfing;
- ☐ board or body surfing;
- ☐ paddle boarding
- ☐ other, specify ....

E. Water based active – individual/group

- ☐ motor boating;
- ☐ water skiing;
- ☐ jet skiing
- ☐ other, specify ....

Where do you mainly participate in these activities?

☛ ***Read the options below and tick as many as apply***

- ☐ Urban, city or town area; *(includes any location in an urban area, including country towns)*
- ☐ Rural or bushland area; *(includes any location in the country, including farms; hills/mountains, rocks)*
- ☐ Ocean or estuaries; *(includes any ocean/beach location, including any nearby islands)*
- ☐ Rivers; *(includes any rivers – metro or rural; canals etc.)*
- ☐ Lakes or dams *(includes any other form of inland water bodies)*
- ☐ Other, specify ....

2. I'm going to read a number of statements about your enjoyment of ***[insert nominated activity/ies]***, and where you recreate, and I want you to tell me how much you agree or disagree with them using the following scale.

☛ ***Explain the scale below and write the nominated number next to each statement.***
VARY THE ORDER YOU READ THE STATEMENTS

1	2	3	4	5
strongly disagree	disagree	neither	agree	strongly agree

I tend to go to the same few places for recreation _____

I am involved in a wide range of recreational activities _____

There are only a few places I'm allowed to go to enjoy my recreational activities _____

More and more people are coming to my favourite places to recreate _____

My recreational activity is being banned in more and more places _____
*(☛ prompt to provide a **general/overall response** if more than one activity and respondent questions it)*

My recreational enjoyment is often lessened by other activities in the same place _____

3. Do you know where your drinking water comes from?

☐ YES ☐ Yes (not certain) ☐ NO ☞ ***Skip to Question 4)***

☞ ***Do not read out options below but tick as many as apply. If they say "scheme water" or "tap water", prompt for specificity.***

Options will change according to the specific context. Following are the examples from the WA Case study.
- ☐ Groundwater *(includes aquifers, specifically named aquifers; anything underground)*
- ☐ Dams *(includes run-off from catchments, from the hills, surface water, weirs, specifically named dams etc.)*
- ☐ Desalination *(this includes sea water)*
- ☐ Rain Water Tanks
- ☐ Other, specify ....
- ☐ Don't Know

4. The water in the dams comes from rain that runs off a large area of land that is called a catchment. To your knowledge, have you ever visited or recreated at a drinking water dam or catchment in WA?

☐ YES ☐ MAYBE ☐ NO ☐ DON'T KNOW

Skip to Question 5

If YES or MAYBE, Where? ☞ ***note as many dams/catchments as nominated***

____

____

In recent years, how often have you visited the dams or catchments?

☞ ***Explain the scale below and note the number nominated***

1	2	3	4
rarely	**occasionally**	**often**	**a lot**

5. How much knowledge do you have about managing the dams and catchments to help protect your drinking water from pollution and contamination?

☞ ***Explain the scale below and note the number nominated***

1	2	3	4	5
no knowledge at all		some knowledge		high level of knowledge

☞ ***If "no knowledge at all" skip to Question 7***

6. What do you think is done to protect the drinking water at dams and catchments from pollution and contamination before it is piped to the treatment plants.

☞ ***Do not give options below, tick those that reflect the answers given, or complete other. Prompt for specificity if answers are vague. An "actual knowledge" score will be calculated here so it's important to get as much information as possible.***

Options will change according to the specific context. Following are the examples from the WA Case study.
- ☐ Keep people away – non specific
- ☐ Keep people and boats out of the water
- ☐ Keep people from going near the water (e.g. 2kms away)
- ☐ Only allowed to picnic or camp in designated areas
- ☐ Only allowed to walk on designated tracks
- ☐ Only allowed to drive on designated roads
- ☐ No motorised activities other than driving on the roads
- ☐ Have to get government permits to hold large scale events
- ☐ Other, specify ....
- ☐ Don't Know

☛ **READ TO THE RESPONDENT** - ***Warn them that there is a bit to listen to and help them through it. It is very important for the future questions that people get this information. Highlighted words are the key points if people ask for repeated information at the next question.***

Following is the policy information from the WA Case study as an example. Information should be no longer and include no greater degree of difficulty either in content or language.

There is considerable evidence that recreation increases the health and wellbeing of society. In WA there are many places available for recreation.

However, the Government currently restricts recreation on our drinking water dams and catchments to provide high quality drinking water. Human contact with the water, and public access within 2kms of the water is not allowed. Activities such as walking and cycling are allowed in the catchments outside the 2km zone. Driving off-road, and any water-based activities such as swimming and boating, are not allowed anywhere in the catchments. This keeps contamination such as human bacteria and pathogens, chemicals, and sediment out of our drinking water, and maintains the natural processes that help to purify the water.

Very little treatment is therefore required to keep our drinking water safe, and the cost to consumers is kept down. If the policy were changed and more recreation was allowed closer to or in the water, the risk of contamination would greatly increase. This would mean increased costs for high level water treatment and recreational management of about $600million. It is estimated that this would increase your water bill by about 15%.

I'm going to ask you a number of questions about how you feel about the policy to protect drinking water, and your thoughts on any possible changes.

7. How much do you agree or disagree with this policy?

☛ ***Explain the scale below and note the number nominated.***

1	2	3	4	5
strongly disagree	disagree	no opinion	agree	strongly agree

Why do you think this?

☛ ***Do not read the options below, but tick as many as apply. Some options may change or be added to in different contexts.***

- ☐ Must protect drinking water
- ☐ Stop contamination
- ☐ Need safe water to drink
- ☐ Cost too much to treat
- ☐ There are other places to go to recreate
- ☐ People are irresponsible
- ☐ Recreation is healthy
- ☐ Not enough places to go for recreation
- ☐ Should have freedom to roam
- ☐ I don't believe recreation pollutes
- ☐ Most people are responsible
- ☐ Other, specify ....

8. How much do you trust the following groups to provide reliable information on catchment protection and drinking water quality?

☞ ***Explain the scale below and write the nominated number next to each agency/group. Vary the order you read the statements***

1	2	3	4	5
no trust at all		**neither**		**complete trust**

Water utility (provide specific name) _____
Health regulator (provide specific name) _____
Water regulator (provide specific name) _____
Scientists _____
Organised Recreation Groups _____

9. I'm going to read a number of statements and I want you to tell me how much you agree or disagree with them using the following scale. Some of them may seem a bit repetitive, but try to ignore this.

☞ ***Explain the scale below and write the nominated number next to each statement. Vary the order you read the statements***

1	2	3	4	5
strongly disagree	**disagree**	**neither**	**agree**	**strongly agree**

I really haven't thought much about managing catchments to protect my drinking water before today. *(Strength of Attitudes)*[7] _____

Being able to recreate in the drinking water catchments is important to me. *(Catchment Protection)* _____

Prevention is better than the cure in protecting our drinking water quality. *(Values & Ethics)* _____

I have complete trust in the authorities to manage our water responsibly. *(Trust in Authorities)* _____

Dams currently used for recreation should not be taken away for drinking water in the future. *(Fair Procedures)* _____

Human concerns should come before the natural environment. *(Values & Ethics)* _____

Compared with other things, protecting the quality of drinking water is not a major issue in my life. *(Strength of Attitudes)* _____

Technology can provide whatever treatment is needed to produce safe drinking water. *(Risk)* _____

When it comes to providing drinking water, the [Water Corporation] is committed to an unbiased process of decision making. *(Fair Procedures)* _____

It is safer to drink water from a clean environment than to rely on artificial treatment. *(Risk)* _____

People should have the right to recreate anywhere that is public land. *(Values & Ethics)* _____

I have very definite thoughts about protecting drinking water. *(Strength of Attitudes)* _____

I insist on being able to recreate anywhere I choose. *(Catchment Protection)* _____

I value having areas that exclude human activity and can provide protection for endangered species. *(Values & Ethics)* _____

---

[7] The words in italics at the end of each statement refer to the attitudinal dimensions being measured. This is to assist in the analysis when developing attitudinal scales. Not all items loaded on the scales of sufficient reliability for analysis in the WA Case Study. However, none are being removed at this stage until the scales have been replicated in a test case study. The words in blue are not for providing to the data collection agency or to be read out by the interviewer. The wording of the statements should not be changed unless it is totally not applicable to the different context.

Protecting our drinking water now is protecting our children's future. *(Values & Ethics)* _____

I have complete trust in the authorities to provide me with safe tap water to drink. *(Trust in Authorities)* _____

When it comes to catchment protection, we should be more concerned about mining and forestry than recreation. *(Risk)* _____

The [Water Corporation] is an important member of the Perth Community. *(Trust in Authorities)* _____

Recreation poses very little risk to drinking water quality. *(Risk)* _____

I believe the authorities should go to any lengths to protect drinking water catchments. *(Catchment Protection)* _____

Protecting drinking water quality is not a personal interest of mine. *(Strength of Attitudes)* _____

I trust the authorities to keep up with the latest water treatment science. *(Trust in Authorities)* _____

It's too risky to allow people to recreate close to, or on the drinking water dams. *(Risk)* _____

I don't believe the [Water Corporation] is part of the Perth Community. *(Trust in Authorities)* _____

People have the right to have good quality drinking water. *(Values & Ethics)* _____

Keeping areas of untouched bushland is important for future generations. *(Values & Ethics)* _____

The [Water Corporation] makes an effort to treat everyone fairly. *(Fair Procedures)* _____

The use of dams for drinking water should always take priority over recreation. *(Catchment Protection)* _____

10. If recreation was allowed [in the 2km restricted zone in drinking water catchments, as well as in and on the water[8]], and additional treatment plants were installed:

☛ ***Explain the scale below and note the number nominated***

How likely do you think it is that something could go wrong with the quality of your drinking water?

1	2	3	4	5
highly unlikely		**neither**		**highly likely**

How serious do you think it would be if something went wrong with the quality of you4r drinking water?

1	2	3	4	5
not at all serious		**serious**		**extremely serious**

What level of control do you think the government authorities would have to stop something going wrong with the quality of your drinking water?

1	2	3	4	5
no control at all		**some control**		**high level of control**

[8] Change according to specific context

☞ ***Read the following to the respondent. The bolded sentences are important pieces of information for the question, so please emphasise.***

WA Case Study Example

As we said previously, the current policy allows no human contact with the water in the dams, and no public access within 2kms of the water.

If any of this policy changed, additional treatment would be needed to provide safe drinking water. This would increase the cost of water to householders by 15%. ***This same level of treatment would be needed whether a small number of extra activities, or a large number of extra activities were allowed.***

Different activities pollute the water in different ways, for example, bacteria and pathogens from human activities, hydro-carbons from motorised vehicles. Some activities are more polluting than others. ***More activities mean more pollutants and these put increasingly greater pressure on the treatment processes. The more the activities, the more the overall risk of something going wrong increases.***

11. I'll read a list of different policy options and I firstly want to know which are ***acceptable*** and which are ***unacceptable*** to you.

☞ ***Always read in the same order as below – increasing levels of risk. Ensure the respondent understands that the more activities, the greater the risk of something going wrong in the treatment process. The risk is increasing from dot point to dot point even though the cost stays the same.***

WA Case Study Example

- *Keep the current policy with no additional cost* ***and no increased treatment risk****.* *Allow off-road activities (such as trail bikes and 4 WD) in the outer catchment, but still no activities in the 2km restricted zone and no contact with water, at added cost* ***and greater risk to your drinking water quality****.*
- *Allow passive activities in the 2km restricted zone (such as walking and picnicking), but no contact with water, at added cost* ***and greater risk to your drinking water quality****.*
- *Allow camping and off-road activities in addition to the passive activities in the 2km restricted zone, but no contact with water, at added cost* ***and greater risk to your drinking water quality****.*
- *Allow passive activities, camping and off-road activities throughout the whole catchment, and fishing, swimming, canoeing and sailing in the water, but no motorised boats, at added cost* ***and greater risk to your drinking water quality****.*
- *Allow passive activities, camping and off-road activities throughout the whole catchment, and fishing, swimming, canoeing, sailing and motorised boats in the water, at added cost* ***and greater risk to your drinking water quality****.*

Of those that are acceptable, which is the most acceptable option to you.

Of those that are unacceptable, which is the most unacceptable option to you.

12. If the drinking water catchment was in a rural location and was the only water body available to that community for recreation, would any of your preferences change?

☐ YES ☐ NOT SURE ☐ NO

Skip to Question 13

If YES or NOT SURE, How would your preferences change?

____

____

13. If the policy was changed to allow greater recreation in the drinking water catchments, who should pay for the additional treatment that would be required?

☞ ***Do not read the options below*** ***but tick the most appropriate answer.***

- ☐ recreators
- ☐ users
- ☐ everyone ***☞ Skip to Question 14***
- ☐ government ***☞ Skip to Question 14***
- ☐ Other, specify ...... ***☞ Skip to Question 14***

If "recreators" or "users": There is considerable evidence that recreation increases the health and well-being of society and so may reduce the government cost of health services. Do you still think that recreators/users should pay?

☐ YES ☐ NOT SURE ☐ NO

Finally I'd like to ask you some questions about yourself and your family

14. Do you belong to any organised recreation group? ☐ YES ☐ NO

15. Would you and your family like to have greater access to drinking water catchments for recreation than you do now?

☐ YES ☐ MAYBE ☐ NO

16. Would you please tell me which of the following categories best describes your age.

18 to 24 years	....	**1**
24 to 39 years	....	**2**
40 to 55 years	....	**3**
56 to 65 years	....	**4**
66 to 75 years	....	**5**
More than 75 years	....	**6**
Refused	....	**99**

17. What describes the unit of people living in your household. (Probe for correct category)

Single adult less than 65 years	....	**1**
Single adult more than 65 years	....	**2**
Two adults – older person less than 65 years	....	**3**
Two adults – older person more than 65 years	....	**4**
Single adult – eldest child less than 18 years	....	**5**
Single adult – eldest child more than 18 years	....	**6**
Two adults – eldest child less than 18 years	....	**7**
Two adults – eldest child more than 18 years	....	**8**
More than two adults – no children	....	**9**
More than two adults – eldest child less than 18 years	....	**10**
More than two adults – eldest child more than 18 years	....	**11**
Refused	....	**99**

18. What country were you born in?

Australia 1
New Zealand/Pacific 2
United Kingdom/Ireland 3
Western Europe/Scandinavia 4
Eastern Europe/Russia 5
USA/Canada 6
Asia/India 7
Middle East 8
Central/South America 9
Africa 10
Refused 99

19. How long have you lived in [Western Australia]?

Less than 5 years 1
5 to 10 years 2
11 to 20 years 3
More than 20 years 4

Note:

Male/Female

Sample Region & Postcode

It is useful to provide the interviewers with additional information (simple and non-technical) to assist with any questions from the respondent. The following proved useful in the WA Case Study.

ADDITIONAL INFORMATION TO ASSIST WITH ANY QUESTIONS

- This is a national research project being conducted for *Water Quality Research Australia.*
- Dams that are not required for drinking water (e.g. irrigation dams) are not subject to the same policy restricting recreation.
- The difference between human contact with the water and animal contact is that humans contaminate the water more due to more sophisticated diets and chemicals in the wide range of medications that people take. Also human activity/impact would increase as the population and demand increased.
- Motorised vehicles and boats contaminate with hydro-carbons and heavy metals.
- Erosion from off-road driving and getting bogged etc. creates sedimentation in the water (which in itself requires treatment) but it also carries pollutants with it in the rain run-off into the dams.
- The average household water bill is $424 and the higher level treatment would add about $60 a year to the average bill.

4.2.3 Data Analysis

After data verification, the data files from the three samples should be joined, while including an identification variable, to form one data base for analysis. Qualitative responses should be coded into key categories and it is suggested that the data be analysed using SPSS[9]. Analytical methods are detailed in the report of the WA Case Study in Appendix 4, and are not overly complex for a suitably qualified social scientist with statistical knowledge.

[9] Statistical Package for Social Sciences

The main analytical processes are:

- Frequencies and means of categorical and multiple responses
- One way analysis of variance to determine statistically significant differences between groups
- Development of multiple item scales after testing for reliability using Cronbach's α co-efficient as an indicator of internal consistency
- Multi-variate analysis using Discriminant Analysis

4.3 Summary of Findings from the WA Case Study

The following provides a summary of the findings from the WA Case Study. Full details can be found in Appendix 4.

The survey involved three different samples with the same questionnaire:

- 604 general community residing at different distances from the public drinking water supply dams and catchments from the Perth Metropolitan area and south to Bunbury and Collie;
- 66 recreators who were members of a range of formal recreation groups who might want to access the drinking water dams and catchments for their recreational activities;
- 28 relevant professionals associated with planning and management of drinking water dams and catchments such as, water managers, health professionals, catchment managers, sustainable tourism and recreation planners, academics and scientists.

The questionnaire was administered by two different processes:

- personal telephone interviews of the random community sample and the recreator sample; and
- web survey of the professional sample.

It is important to note that the emphasis of this survey was on drinking water quality and its protection and the preference by different stakeholder groups for different levels of access for recreation in drinking water catchments. As such it does not deal with recreation planning *per se*.

It was evident that most people participated in some form of outdoor recreation in the survey region with only 16% of the community sample saying they did not. A wide range of land- and water-based activities, both passive and active, were nominated by respondents when asked. These activities also occurred in a wide range of locations. **However, 29% of the community sample and 80% of the recreator sample indicated some discontent with opportunities to enjoy their chosen activities.**

There was poor awareness of the range of drinking water sources by the community sample. Desalination was a lesser known source by both the community and recreator samples. Dams was the best known source with almost a quarter of the community sample, 86% of the recreators, and 96% of the professionals having visited or recreated at a drinking water dam or catchment in WA. However, when asked to nominate the dams or catchments they had visited, respondents did not necessarily discriminate between drinking water and irrigation dams. **While almost half the community sample said they *rarely* visited the dams or catchments, 14% said they visited *often* or *a lot*. About half the other two samples said they visited the catchments *often* or *a lot*, though a number of professionals did so in the course of their work.**

Respondents were asked to both rate their perceived knowledge about managing dams and catchments to help protect drinking water from contamination, and to also say specifically what they knew. From the second question an *actual* knowledge score was derived. There was only a moderate correlation between what people *thought they knew* and what they *actually knew*. **The community had poor knowledge of drinking water catchment protection, while the recreators and professionals had a significantly better knowledge than the community, but not different from each other.**

Respondents were provided with a brief description of the current policy of restricted recreational access to drinking water catchments, which included basic reasons why and the implications for treatment and additional cost if greater recreational access was allowed. **When asked to rate**

agreement or disagreement with the current policy, 86% of the community and 57% of the professionals agreed with it, while 71% of recreators disagreed. A wide variety of reasons for agreement and disagreement with the policy were provided unprompted. The major reasons provided by the community for agreement were *must protect the drinking water* and *stop contamination of drinking water*. The main reasons provided by the professionals for agreement were *stop contamination of drinking water* and *prevention is better than the cure*. The main reasons for professionals' disagreement with the policy were *some activities could be allowed* and *it is not policed*. The main reasons for the recreators' disagreement with the policy were *I don't believe recreation pollutes* and *recreation is allowed in other parts of Australia/the world*. Interestingly, **only about a fifth of the community sample nominated *increased cost* as a reason for agreement with the policy.** This is contrary to frequent assumptions that the community will base its decisions primarily on monetary issues.

The rating of trust in particular relevant authorities and organised recreation groups to provide reliable information about drinking water catchment protection showed that quite high proportions of the community and the recreators were unsure about whether they trusted the different groups. This included high uncertainly by the recreators about whether they trusted organised recreation groups as reliable information providers. **Scientists and the Department of Health were most trusted by the community and the recreators to provide reliable information, and the community was generally trusting of the Water Corporation and the Department of Water.**

A number of attitudinal scales were devised from responses to attitudinal statements and a number of questions. These were tested and proved to be reliable measures[10].

Recreational Enjoyment: Scores on the recreational enjoyment scale indicated general enjoyment of activities by the community and professionals with the recreators enjoying their activities statistically significantly less than the other two samples.

Trust in Authorities: Scores on the scale about trust in authorities to provide safe drinking water quality showed the recreators to be least trusting of authorities, with the community being more trusting than recreators but less trusting than the professionals.

Catchment Protection: Scores on the scale about catchment protection and allowing recreation in drinking water catchments indicated agreement with protecting drinking water catchments by both the community and the professionals while the recreators were less in favour than the other two samples.

Precautionary Ethics: Scores on the scale of precautionary ethical statements (i.e. human and environmental caution in both the present and the future) showed the community and the professionals to be more conservative than the recreators, though this was more in degree than opposing views.

Fair Procedures: Scores on the scale in regard to the Water Corporation's decision making processes in providing drinking water showed that the community and the professionals considered the processes to be generally fair, with the recreators considering them to be significantly less fair.

Strength of Attitudes: Scores on the scale indicating strength of attitudes about drinking water and its protection showed the community and the recreators to have generally uncertain attitudes with the professionals moderately strong attitudes being statistically significantly stronger from the other two samples.

Catchment Risk: Scores on the scale about risk associated with catchment management and recreation showed the community and the professionals to be quite highly risk averse and significantly different from the recreators who were generally risk tolerant.

System Risk: A *system risk* scale was calculated from three questions about perceived likelihood and seriousness of a possible drinking water quality incident if recreation was allowed in the catchments and the perceived degree of control authorities would have to stop it from occurring. Scores on this scale again indicated significantly higher concerns about risk by the community and professionals than the recreators.

Respondents were asked to consider a range of possible policy options which involved increasing access by land- and water-based recreational activities in drinking water catchments with the

[10] Scales comprised of a number of responses to statements and questions provide more reliable measures than single items.

consequences of one increased treatment cost for all options, but increasing risk associated with the treatment processes with increasing activities and contact with water. They were asked to firstly nominate those options that were acceptable or unacceptable and then to say which option was most acceptable and which was most unacceptable.

The community sample strongly supported retention of the current policy with two-thirds nominating it as the *most acceptable* option and a further 22% nominating it as *acceptable*. While a similar proportion of professionals nominated it *most acceptable*, only 10% of the recreators did so.

However, **45% of the community and 39% of the professionals also considered passive activities in the 2km restricted zone to be either *acceptable* or *most acceptable*. Seventy-nine percent of recreators supported this policy option.**

Also, 38% of the community sample and 18% of the professionals considered greater off-road access to the outer catchment to be either *acceptable* or *most acceptable*. Fifty-six percent of the recreators supported this policy option.

There was much less support by all three samples for these latter two policy options when combined to allow a considerable increase in land-based activities in the catchment.

There was very little support from the community and professional samples for either option that allowed contact with water, while 41% of recreators supported non-motorised water activities.

Discriminant analyses were conducted to determine the key variables that discriminated acceptability and unacceptability of the current policy and the other two options that were somewhat supported.

Results showed that the respondents who **considered the current policy to be most acceptable:**

- were more likely to be **risk averse** when considering the risk to drinking water from recreation
- were more likely to think that the **Water Corporation used fair procedures in their decision making**
- were inclined to consider that **a water quality incident would be more likely, more serious and with lesser control by the authorities to stop something from going wrong** with water quality if the catchments were opened up to recreation
- were more likely to **trust the Water Corporation to provide reliable information** on catchment protection and drinking water quality
- had a **lower actual knowledge** score about what's done to protect drinking water dams and catchments from pollution and contamination
- had **more conservative precautionary ethics**
- had **slightly less central, or definite, attitudes associated with drinking water and its protection**.

The acceptance groups of the other two policy options were not as well discriminated as for the current policy, although still statistically significant. For both policy options, only two variables emerged and results showed that the respondents who ***considered each alternative policy to be most acceptable*:**

- had **lesser conservative precautionary ethics**; and
- were inclined to consider that **a water quality incident would be less likely, less serious and with some control by the authorities to stop something from going wrong** with water quality if the catchments were opened up to recreation.

There were a number of statistically significant differences in responses to key questions in the survey by respondents of different demographics: gender; age; location and membership of organised recreational groups.

In regard to the latter demographic difference, responses from people in the community sample who advised that they were members of organised recreation groups (but not sporting clubs or groups) were compared with those of the recreator sample. **Mean responses to all key questions by the community formal recreators mirrored those of the whole community sample and were significantly different from the mean responses of the recreator sample.** Given this, it would appear that the people in the recreator sample should be considered as "champions" of recreation and looking to bettering future recreational opportunities in the light of current and future demands. They generally saw current policy as being too conservative compared with practices elsewhere and their considerations of the risks involved.

Two key observations are made in the conclusions in this report. The first relates to the potential future demand for recreation in the region given population growth, and also the degree to which quite a large proportion of the community as well as the recreator sample are finding increasing difficulty in gaining enjoyment from their chosen activities in general.

- *Given current and potential future demands for recreation facilities and the level of current discontent in the ability to enjoy activities, it would seem that urgent attention needs to be given to recreational planning and management in the region as a whole.*

The second is in relation to the community's very low *actua*l knowledge about catchment protection for drinking water and also poor knowledge about the range of sources for their drinking water. It also relates to the quite high proportions of community and recreators who were uncertain about their trust in the various authorities to provide reliable information.

- *If future informed debate is to be conducted on catchment protection, these issues of knowledge and its provision will need consideration. Note should be taken of the trusted information providers for any communication or information program, especially the trust in scientists and the Department of Health displayed by all groups.*

From the findings of the survey, it is apparent that **the next, risk assessment stage of the project should initially consider the current policy option as well as the other two, somewhat supported, options in light of future population growth, potential drinking water sources, and climate change.** The risk assessment will involve the range of professional disciplines represented in the survey, including catchment managers, recreation and tourism professionals and planners, as well as the traditional water and health managers. Key findings from this survey will be incorporated in the risk assessment, including the apparent requirement for future regional recreation planning and management, and community concerns about risk and precautionary management.

5. RISK-BENEFIT ASSESSMENT PHASE

The aim of this phase of the project is to utilise the results of the survey, in particular the identification of acceptable and somewhat acceptable policies, and conduct a risk assessment to determine the optimal policy for implementation. The risk assessment involves the range of professional expertise as has already participated in the project to this phase. Also, as shown in the previous phases, it is important to include the assessment of benefits in the identification of the optimal policy.

In the original proposal it was thought that the risk assessment model developed by the CRC for Water Quality and Treatment would form the basis of the methodology chosen for this final phase of this project. However it soon became evident that, due to two factors that emerged in the preceding phases of the WA Case Study, an alternative method needed to be developed.

Firstly, while there was a consensus among the professionals that the current policy of catchment management had "worked" for over a hundred years, there was far from detailed agreement as to the science of how and why it had worked. In fact, the science of the adverse effects of recreation on water quality, while having some considerable support, was doubted by others who observed recreation occurring in water catchments elsewhere in Australia and overseas without any apparent ill effects. As such, the additional treatment costs that the Water Corporation would implement with any change of policy in the inner catchment were regarded as over cautious by some in the early stages of the research. These queries about the actual reality of the risk applied to all three policies that emerged from the survey analysis for investigation in this final phase of the project.

Secondly, there were obvious concerns that there were significant benefits being foregone in restricting access for recreation in the catchments. There were clear indications from the survey that there was growing dissatisfaction with abilities to fully enjoy recreation activities of choice. As with the water quality science mentioned above, the extent of these benefits are difficult to precisely define, but the literature is clear that those involved with more passive activities in forests have been shown to be more environmentally knowledgeable than other recreators, and are prepared to become more actively involved in environmental protection, thus providing a potential resource for good management of the RPZ. (Teisl and O'Brien, 2003).

It was evident that these arguments had been persisting for some time and were at the heart of the conflict between drinking water protection and recreational access to catchments in WA. And from literature reviews and discussions with the project Steering Committee, these arguments were not confined to the WA Case Study, but the nature of the conflict seemed to be similar throughout Australia.

Therefore, a technique which could combine informed opinions among water, health, planning, environmental, recreation and tourism professionals was required that could deal with both risks and benefits (Finucane et al, 2000) within a framework of highly uncertain science and socio-economic analysis. Therefore, a simple risk benefit hierarchical or multi-attribute utility (MAU) approach was chosen as the best way to promote discussion, and provide a transparent outcome for all professional interests within a reasonable time. These simple linear decision approaches have proven to be pragmatic and successful in a variety of situations where professional judgments are needed on a situation by situation basis. (Dawes and Corrigan, 1974; Dawes, 1979; Edwards,1979; Gardner and Edwards, 1975; Hogarth and Natalia, 2007; Weiss, Weiss and Edwards, 2010).

The methodology first identifies the range of risks and benefits associated with the implementation of the policy or proposed policy(ies), and these are then categorised. Participants can then rate the relative importance of the components within each category, and the relative importance of the categories as a whole, for both the risks and benefits associated with each of the policies being assessed (Wright and Goodwin, 1999). A different multi-attribute hierarchy can be created for each policy, which enables the differing issues arising for each policy to be assessed in terms of their significance. In this way, the important risks and benefits that will need to be dealt with if the policy is to be adopted can be readily seen.

While such a risk-benefit hierarchy is simple and its calculation visible to all participants, it does have one important assumption that the different categories (and their components) are statistically independent. On investigation of the results from the rating exercise in the WA Case Study (as shown in Appendix 5), it was evident that the importance ratings for the risks and benefits were not independent. Despite the relatively low number of participants in this study, there were highly significant correlations between some of the risks, between some of the benefits, and between some of the benefits and risks.

A version of cross impact assessment was devised by Enzer and colleagues (e.g. Enzer, 1984; Enzer and Alter, 1978) in which these relationships are acknowledged. This technique can provide sophisticated analyses of the overall acceptability of a policy when the interactions between its components are considered. Optimisation statistical analyses can then be employed to refine policies to achieve the best outcomes. The technique works particularly well when there are objective or measurable indicators of categories or components of the categories. Unfortunately, due to subjective nature of this task, and the requirement to develop a consensual approach to risk assessment, a literal application of the cross impact approach was deemed impracticable.

For this reason, a simplified, small group, cross impact analysis was developed to follow on from the MAU importance ratings. This technique does two things. It looks at the combined (added) importance of each pair of categories and then multiplies them by the strength of their relationships. This process results in the development of a policy matrix which shows the relative influence of each risk and benefit category in the overall policy assessment in the light of its relationship with other categories. This gives a clear indication of the significant issues that will face each policy if implemented. By conducting the cross impact analysis as a small group, some elements of a Group Delphi approach (e.g. Kendall, 1977; Preble, 1983) were used to encourage group consensus (Biloslavo and Dolinsek, 2010; Fu and Yang, 2010; Yearwood and Stranieri, 2010).

As this was a new[11] method devised during the WA Case Study, a number of improvements were identified in its conduct. These improvements have been incorporated in the following method description.

5.1 Risk-Benefit Assessment Method

The risk/benefit assessment should be conducted in a three stage process.

- A full day workshop to develop the risks and benefits of each policy and organise into categories.
- Individual MAU rating of the risk and benefit categories and components conducted by email.
- A full day workshop to complete the cross-impact assessment.

Full day workshops are suggested here as there were three policies for assessment in the WA Case Study. If there are fewer, lesser time may be possible. However, it is suggested that no more than three policies could be assessed in any workshop.

An overall objective of the risk-benefit assessment should be developed as well as Guiding Principles. These should reflect both the overall objective of the project, as well as the findings of the survey phase. That is, the key philosophies and principles that arise from the analysis of the data from the survey of stakeholders should guide the assessment of risks and benefits of the policy options.

The context for the assessment needs to be clearly articulated. That is, what will the future look like in which the policies will have to function. For example, in the WA Case Study, the current policy of drinking water catchment protection was unequivocally supported in the survey, but with some support for two other policies that involved slight increases in recreational access to the catchments. What was also clear from consideration of the survey findings was that recreational demand would increase with expected population increases, and that apparent dissatisfaction with the current lack of sufficient recreation facilities would also increase and thus put greater pressure on existing facilities. Subsequently, demands for greater access to the drinking water catchments would also increase. Therefore, a hypothetical scenario was developed thirty to forty years into the future of Perth and the South West, to encourage the professionals to think further than "what's now". While the current policy was right for now, there was a need to consider the workability of both the current policy and the two somewhat acceptable policies into the future, especially as pro-recreation lobbies would continue.

Even though other contexts may need to assess the most workable policy for current times, there is still necessary to consider the future if a lasting result is to be achieved. The number of years into the future for consideration will vary from context to context, but it is suggested that the minimum be 15 to 20 years. The development of the future scenario should not concentrate on "everything factual" but be designed to extend current thinking.

[11] "New" in terms of the combination of techniques, but the individual aspects of the method have been tried and tested by the researchers over many years.

The following describes the methodology in detail. Examples are used from the WA Case Study to illustrate the different tasks and stages.

5.1.1 Invitation of Participants

A sub-set of the professional sample from the survey should be selected so that a range of professional backgrounds/expertise participate in this stage of the project. This should include not only water and health professionals, but also tourism, recreation, planning, and environmental and forestry management. As approximately 12 participants are required to participate, sufficient numbers need to be invited to allow for refusals and for drop outs. This will depend on the local context, but it is suggested that about 18 be invited with the assumption that about 14 will accept and 12 will attend the first workshop. However, as in previous phases of this work, it is more important to gain the participation of a *range* of disciplines and organisations associated with the issue, than to have exactly the targeted number.

The professionals should be sent a copy of the survey report and invited to participate in the three stage risk/benefit assessment process. Email invitations are appropriate. The invitation should outline the three stages and the dates these will occur while advising that acceptance of the invitation is a commitment to participate in all three stages. The importance of this should be stressed as it is very difficult for a replacement to come in later in the process without having had the benefit of the previous discussions.

5.1.2 Developing the Hierarchical Framework

The following section outlines the development of the hierarchical framework illustrated by examples from the WA Case Study. It involves a full day's workshop and out of session individual work. The full framework development from the WA Case Study can be found in Appendix 5.

5.1.2.1 Workshop 1 - Identification of the Risks and Benefits

The aim of the first workshop is to identify the risks and benefits associated with the implementation of the preferred policy options identified in the Survey Phase and to structure them into categories. The following outlines the format and method for the first workshop. Time allocations are suggested only. Refreshment breaks and lunch should be provided when deemed appropriate.

Background briefing (1.5 hours)

Some participants will have been involved in both the Scoping and the Survey Phases of the project, while others will only have been respondents to the survey. The start of the workshop should ensure that all participants understand the aim of the project, and are briefed on the findings of both the previous phases. They should also be given the opportunity to ask any questions of clarification in regard to the previous phases.

Agreement with the Assessment Objective, Guiding Principles and Future Scenario (1 hour)

As noted above, the Objective of the risk/benefit assessment and the Guiding Principles should be based on the survey findings. The future scenario should not be purely fact, but designed to extend thinking beyond current times. All these should be presented to the workshop participants who should be given the opportunity to make minor changes, while respecting the findings of the survey. Following are examples taken from the WA Case Study.

POLICY OBJECTIVE FOR THE RISK-BENEFIT ASSESSMENT - WA CASE STUDY

> *To improve the health and well-being of the Perth and South West community by providing a high quality drinking water supply from drinking water catchments while supporting catchment management that provides the best practicable level of recreation opportunities in the light of competing land and water uses.*

GUIDING PRINCIPLES FOR THE RISK-BENEFIT ASSESSMENT – WA CASE STUDY

> *The risk-benefit assessment should be guided by the following principles while considering future climatic and population trends.*

1. *A precautionary approach should be maintained, with special consideration paid to the maintenance of drinking water quality and ecosystem protection for future generations.*
2. *A precautionary approach should apply to assessments of catchment risk and system risk and include, as far as practicable:*
   2.1. *the avoidance of contamination rather than a reliance on artificial treatments;*
   2.2. *a system which minimises the likelihood of a significant water quality incident.*
3. *Assessments should include evidence of procedurally-just decision making, and distributive justice for the overall well-being of all groups.*
4. *Adaptive management should be applied to achieve fit-for-purpose water quality and promote recreational choice and enjoyment for the metropolitan and South West community*

HYPOTHETICAL FUTURE FOR PERTH & THE SOUTH-WEST: 2030 – 2040

	2030	***2040***
Perth Population Growth	*33%*	*48%*

Increasing numbers of retirees will put greater demand on the south-west infrastructure and landscape. Without careful management, the predicted population boom in the South West and subsequent economic growth in the region will contaminate drinking water sources.

Recreation demand on all catchments

(people per annum – based on figures from the survey)

	2030	***2040***
Casual demand	*1,134,000*	*1,258,000*
Regular demand (visit often or very often)	*300,500*	*338,300*
Dissatisfied recreators (any visitation frequency)	*248,000*	*340,000*

There will be increased demand on the Bibbulmun & Munda Biddi tracks from growing WA population and increased national and international tourism. There will be a need for better planning for recreation to obtain water allocations for the required water levels for the different activities.

Implications of Climate Change

Dams and catchments

Climate change will mean changing emphasis on the traditional sources of drinking water (surface water dams and groundwater aquifers) and increasing emphasis on manufactured water (recycled and desalinated). Manufactured water will be stored in existing dams and aquifers. This will mean that some dams will increase in importance and security requirements while others will decrease. Some dams will be abandoned as potential drinking water sources. Some water bodies will decrease in volume and cause problems for water-based recreation.

Fit-for-purpose water quality

There will be more bushfires and more extreme storms resulting in increased criticality of keeping drinking water quality protected.

It is predicted that demand for water will outstrip supply in farmland stretching from Harvey to Pemberton by 2030. The importance of irrigation dams for food production and other development will increase, and the maintenance of acceptable water quality will become an issue for some recreational activities (e.g. motorised water-based activities).

There will be increasing requirements to manage for recreational water quality.

Recreation

Increasing temperatures will mean an increased demand for water-based recreation, as well as a move to more in-door activities.

Competing Landuses

There will be increasing pressure from competing landuses. Population growth will require urban development and to decrease the north-south coastal sprawl, there will be pressure to develop to the east. Increased national and international tourism industry will demand the ability to provide resort accommodation overlooking water and near the walking and cycling tracks. Forestry and mining will continue to be competing landuses.

Trends in Drinking Water Treatment

There will be a move towards national regulation of drinking water standards rather than current guidelines. With the more extreme events associated with climate change, filtration will become more common. There will be growing international pressure for more stringent water treatment, such as filtration of all water at all times.

Energy use

There will be increasing requirement to minimise greenhouse gas emissions from increased energy use for higher level water treatment, and the manufacture and transport of drinking water supplies.

Community Health

Health issues associated with life-style and climate change will require a greater emphasis on preventative health programs.

Presentation of Policies for Assessment (15 minutes)

Participants will be familiar with the policies for assessment having responded to the survey. However, these should again be provided in simple outline form and questions of clarification encouraged. It is not the intention to provide detailed information, but only the key aspects that have implications for recreation. The WA Case Study examples follow.

Current Policy

- *No public access to the 2km Reservoir Protection Zones (RPZ).*
- *Passive activities allowed in the outer catchments.*
- *Driving off-road is not allowed anywhere in the catchments.*
- *No water-based activities are allowed on the dams or in the 2km RPZ of the feeder streams.*

Alternative Policy #1

The Current Policy with the addition of:

- *allow passive activities in the 2km RPZ.*

Driving off-road, and any water-based activities are still not allowed.

Alternative Policy #2

Current Policy with the addition of:

- *allow off-road activities (such as trail bikes and 4 WD) in the outer catchment.*

Still no public access to the 2km RPZ and no contact with water.

Individual Work (30 minutes)

Participants are firstly provided with about thirty minutes to spend on their own individual assessments of risks and benefits of implementing the three policies under the future scenario from the point of view of their particular areas of expertise. This is then used as reference for the group discussion.

Group Work (2 hours)

Participants should be assigned to one of two groups, ensuring a similar mix of expertise in each group. They will then be required to discuss and amalgamate their individual points of view and produce group considerations of risks and benefits associated with the implementation of the policies in the future. The groups should then structure the individual risks and benefits into categories and place these on the wall for feedback and discussion. The facilitators should ensure as much detail as possible is provided. For example, in the WA Case Study, public health benefits and risks were stated generally, but the groups need to be more specific and provide detailed risks and benefits.

Final Plenary Session (45 minutes)

The final session of the day's workshop allows the groups to explain their output. This is best done through one group presenting in detail for a policy and the other group providing any differences they may have. Discussion should allow the facilitators/researchers to gain a detailed understanding of the categories and their component risks and benefits so they can structure the next stage in an informed fashion.

The final session should also allow participants to provide any insights they may have had as a result of the day's discussions. Anonymous Feedback Forms should also be provided.

5.1.2.2 Individual MAU Rating of the Risks and Benefits

After the workshop the researchers should combine and synthesise the two groups' output to provide the hierarchical framework for further consideration. That is, develop one set of risk categories and their components and one set of benefits categories and their components for each policy option being assessed. Tables 3 and 4 below show examples of the hierarchical framework for the current policy in the WA Case Study.

Table 3. Risks Associated with the Implementation of the Current WA Policy in the Future

Risk Category	Risk Component
Lost Opportunities	General recreational demand not met
	Nature based recreation not met
	Eco-tourism opportunities lost
	Reduced enjoyment through overload of insufficient facilities
	Lost opportunity to educate the public about catchments
	Lost opportunity for eco-education activities
Implementation & Policing	False sense of security of drinking water safety
	Overload on outer catchment and facilities
	Inability to police and enforce the policy
	It's "fun to disobey"
	Public dismissal of the policy
Cost	High cost of policing the policy
	Legal risks associated with prohibited access
	Loss of near-water land for other uses
	Inefficient use of land for water supply compared with manufactured sources
	Opportunity cost for alternative land uses

Table 4. Benefits Associated with the Implementation of the Current WA Policy in the Future

Benefit Category	Benefit Component
Public Health	Low human health risks
	Human health benefits from recreation available outside the 2km RPZ
	Provides source diversity to ensure supply under conditions of uncertainty
Water Quality	High quality water with little treatment
	Lack of reliance on advanced water treatment
	Avoids possible contaminants (prevention better than the cure)
	Reduced fire risks
	2 km RPZ filters the water
	Reduced access to feeder streams with 2km RPZ
Environmental Health	Reduced risk of dieback
	Protected ecology and maintenance of biodiversity in 2km RPZ
	Minimised erosion
	Reduced risk of introduced species and diseases
	Low carbon footprint
	Scientific benefit from intact ecosystems
Cost	More affordable water
	Low cost system implementation currently
	Public liability reduced
	Simplified management
	Convenience of maintenance of status quo
Social & Cultural Issues	Public trust in safe drinking water supplies
	Indigenous heritage protected
	Reduced recreational competition and conflict
	One policy for all in the RPZ
	Intergenerational equity of good quality water supply and environmental existence

As much consistency in wording as possible should occur across the policies. Also, the groups tend to not repeat some risks and benefits that are applicable to all policies, so the researchers should include them in all policies in their synthesis. The risks and benefits for each policy options should then be structured into Excel Worksheets[12] with instructions, and emailed to the workshop participants for completion and return by a certain date. The instructions are as follows (as appeared in the Excel Workbook in the WA Case Study).

PROCESS FOR POLICY RISKS

[12] All cells should be locked except those that require participant input.

Complete each category separately. Read all the risks in the category and while ***thinking about the consequences and the likelihood of each,*** **rate its importance** in the assessment of the policy.

Rating Process: Assign the value "10" to the ***risk of least importance in the category*** and type it in the adjacent coloured cell. Then rate the other risks in the category ***relative to the one of least importance***. That is, if another is twice as important, assign the value of "20". If it is 5 times as important, assign the value "50". Do not go higher than 10 times as important (100). You can have risks of equal importance. See the example below.

Complete a similar process for the **RISK CATEGORIES** as a whole. Consider the ***consequences and likelihood of all the risks in each category***, and rate the importance of each category as a whole relative to the others.

Example for Risk Category B – Implementation and Policing

B	IMPLEMENTATION AND POLICING		*While thinking about the consequences and likelihood of each risk*
	False sense of security of drinking water safety	10	Of **equal least importance** when assessing this policy
	Overload on outer catchment and facilities	10	
	Inability to police and enforce the policy	70	**7 times more important** when assessing this policy
	It's "fun to disobey"	50	**5 times more important** when assessing this policy
	Public dismissal of the policy	20	**Twice as important** when assessing this policy

PROCESS FOR POLICY BENEFITS

Complete each category separately. Read all the benefits in the category and then while *thinking about the outcomes and the likelihood of each,* **rate the importance** of each in the assessment of the policy.

Rating Process: Same as above -- assign the value "10" to the *benefit of least importance* and type it in the adjacent coloured cell. Then rate the other benefits in the category *relative to the one of least importance*. See the example below.

Complete a similar process for all the BENEFIT CATEGORIES as a whole. Consider the outcomes and likelihood of all the benefits in each category, and rate the importance of each category as a whole relative to the others.

Example for Benefit Category A – Public Health

A	PUBLIC HEALTH		*While thinking about the outcomes and likelihood of each benefit*
	Low human health risks	100	**10 times more important** when assessing this policy
	Human health benefits from recreation available outside the 2km RPZ	10	**least important** when assessing this policy
	Provides source diversity to ensure supply under conditions of uncertainty	60	**6 times more important** when assessing this policy

5.1.2.3 Analysing the MAU ratings

It is highly likely that there will be considerable variation in the extremity of rating responses between participants. To account for this, the ratings should be standardised so that all ratings within a category sum to 100, and the category ratings also sum to 100.

For example, suppose there are 4 benefit components in a benefit category, the equation to standardise the ratings of each component is as follows:

$$stbr1 = (br1/(br1+br2+br3+br4))*100$$

where *stbr1* is *standardised benefit component rating 1,* and *br1 ...... br4* are *benefit component ratings 1 to 4.*

Similarly, standardise the category ratings.

Figures 3 and 4 below show an example of participant ratings before and after standardisation. Note the difference in intensity between the original ratings by the eleven participants.

RISKS											
LOST OPPORTUNITIES											
General recreational demand not met	80	30	40	100	10	80	30	10	10	20	10
Nature based recreation not met	60	20	50	80	80	50	70	10	10	20	10
Eco-tourism opportunities lost	60	30	20	80	60	50	20	10	10	20	10
Reduced enjoyment through overload of insufficient facilities	50	20	10	10	80	80	20	20	10	20	10
Lost opportunity to educate the public about catchments	10	10	30	80	30	10	40	10	20	10	10
Lost opportunity for eco-education activities	60	10	10	50	60	50	10	10	10	10	10
Sum	320	120	160	400	320	320	190	70	70	100	60

Figure 3: Participant Ratings of Risk Components associated with the Implementation of the Current WA Policy in the Future before Standardisation

RISKS											
LOST OPPORTUNITIES											
General recreational demand not met	25.0	25.0	25.0	25.0	3.1	25.0	15.8	14.3	14.3	20.0	16.7
Nature based recreation not met	18.8	16.7	31.3	20.0	25.0	15.6	36.8	14.3	14.3	20.0	16.7
Eco-tourism opportunities lost	18.8	25.0	12.5	20.0	18.8	15.6	10.5	14.3	14.3	20.0	16.7
Reduced enjoyment through overload of insufficient facilities	15.6	16.7	6.3	2.5	25.0	25.0	10.5	28.6	14.3	20.0	16.7
Lost opportunity to educate the public about catchments	3.1	8.3	18.8	20.0	9.4	3.1	21.1	14.3	28.6	10.0	16.7
Lost opportunity for eco-education activities	18.8	8.3	6.3	12.5	18.8	15.6	5.3	14.3	14.3	10.0	16.7
Sum	100	100	100	100	100	100	100	100	100	100	100

Figure 4: Participant Ratings of Risk Components associated with the Implementation of the Current WA Policy in the Future after Standardisation

After the component and category importance ratings are standardised, the component standardised importance should be multiplied by the category standardised importance to provide overall *importance weights* for both the risks and the benefits. These then should also standardised so that all the weightings (risks and benefits components) sum to 100 and ordered in terms of standardised importance weights. For example, Table 5 shows the standardised ratings of the risk and benefit categories in the WA Case Study, and Table 6 shows the standardised weightings for the risk and benefit components associated with implementing the current WA policy under the future scenario. Table 6 shows the risk and benefit components of greatest importance in the implementation of the current policy into the future and it can be seen that there are more risks than benefits at the top of the table.

Table 5. Standardised Ratings of Risk (black) and Benefit (grey) Categories associated with the Implementation of the Current WA Policy in the Future

Categories		Standardised Ratings
RISKS	LOST OPPORTUNITIES	37.89
	IMPLEMENTATION AND POLICING	36.84
	COST	25.28
BENEFITS	WATER QUALITY	27.84
	PUBLIC HEALTH	27.08
	ENVIRONMENTAL HEALTH	20.60
	SOCIAL & CULTURAL ISSUES	13.06
	COST	11.42

Table 6. Standardised Weightings of Risks (black) and Benefits (grey) associated with the Implementation of the Current WA Policy in the Future

Risks and Benefits	Standardised Weight
Low human health risks	6.41
Inability to police and enforce the policy	5.74
Provides source diversity to ensure supply under conditions of uncertainty	4.28
High cost of policing the policy	3.97
Overload on outer catchment and facilities	3.96
Nature based recreation not met	3.95
False sense of security of drinking water safety	3.69
High quality water with little treatment	3.60
General recreational demand not met	3.60
Eco-tourism opportunities lost	3.21
Reduced enjoyment through overload of insufficient facilities	3.12
Public dismissal of the policy	2.91
Human health benefits from recreation available outside the 2km RPZ	2.85
Avoids possible contaminants (prevention better than the cure)	2.78
Lost opportunity to educate the public about catchments	2.64
Opportunity cost for alternative land uses	2.47
Lost opportunity for eco-education activities	2.42
Lack of reliance on advanced water treatment	2.29
Loss of near-water land for other uses	2.16
Reduced access to feeder streams with 2km RPZ	2.15

Table 6 cont.

Risks and Benefits	Standardised Weight
Legal risks associated with prohibited access	2.15
It's "fun to disobey"	2.12
Reduced risk of dieback	1.98
Protected ecology and maintenance of biodiversity in 2km RPZ	1.94
Public trust in safe drinking water supplies	1.90
Inefficient use of land for water supply compared with manufactured sources	1.89
Minimised erosion	1.84
2 km RPZ filters the water	1.79
Scientific benefit from intact ecosystems	1.73
More affordable water	1.71
Intergenerational equity of good quality water supply and environmental existence	1.65
Reduced risk of introduced species and diseases	1.52
Low cost system implementation currently	1.41
Reduced fire risks	1.31
Low carbon footprint	1.30
One policy for all in the RPZ	1.02
Indigenous heritage protected	0.99
Reduced recreational competition and conflict	0.97
Simplified management	0.93
Public liability reduced	0.92
Convenience of maintenance of status quo	0.75

The relative importance weights of the components are then used for reference in the consideration of the cross-impact relationships of the categories in the next stage of the risk-benefit assessment.

5.1.3 Cross Impact Assessment

As explained previously, correlation matrices of the standardised importance ratings of risks and benefits for each policy were examined and these showed a considerable number of statistically significant correlations between risk and benefit components. Therefore, it was decided to work at the category level while noting the components within the categories.

5.1.3.1 Development of the Cross Impact Matrices

Preliminary matrices of risk and benefit categories for each policy should be prepared as shown in Figure 5 below. The risk (red) and benefit (green) categories for the policy form the basis of the matrix. The impact of the combination of each pair of categories are assessed using two measurements.

- The ***combined importances*** of each pair of categories
- The ***strength of relationship*** between each pair of categories

The first (***combined importances***) is shown in the matrix as a result of the MAU rating exercise. The black numbers in each cell in Figure 5 are the ***summed standardised importance ratings*** of each pair of categories.

The second (***strength of relationship***) will be determined by the participants at the second workshop. In the meantime, these are shown as empty purple brackets in the cells, and a legend in purple at the side of the matrix displays the ***strength scale*** to be used in determining this.

The ***combined importance*** **will then be multiplied by the** ***strength of relationship*** in each cell to provide an overall impact score, and entered in the appropriate yellow shaded cells in the matrix.

5.1.3.2 Workshop 2 - Completing the Cross Impact Assessment

The following outlines the procedure at Workshop 2 where the professional participants complete the cross impact assessment of each policy option.

Participants should be asked to bring their MAU ratings to the workshop for individual reference. Copies of the matrix for each policy, as described above, should be provided to the participants. As in Workshop 1, participants should be assigned to one of two groups which provide a mix of disciplines/expertise in each. If the groups in Workshop 1 "worked" then the same group membership can be maintained in Workshop 2. However, the facilitators may decide to change membership as a result of observation of interactions at Workshop 1. The following is a suggested format and timing for the different tasks at the workshop. Refreshments and lunch should be provided at appropriate times.

Plenary Session (45 minutes)

The purpose of the plenary session is to provide feedback on the results of the MAU rating process and how this provides the basis for the cross impact assessment, as well as outlining instructions for the workshop tasks.

The feedback presentation should show the participants the range of MAU rating responses while ensuring anonymity. It is important that they see the extent of the differences in the ways they each viewed the risks and benefits of the policy options. Feedback should also include the lists of risk and benefit components in decreasing order of importance according to the *standardised weights* for each policy option.

(strength of relationship) combined importance

	LOST OPPORTUNITIES	IMPLEMENTATION AND POLICING	COST	PUBLIC HEALTH	WATER QUALITY	ENVIRONMENTAL HEALTH	COST	SOCIAL & CULTURAL ISSUES
LOST OPPORTUNITIES		( ) 75	( ) 63	( ) 65	( ) 66	( ) 58	( ) 49	( ) 40
IMPLEMENTATION AND POLICING	strength x importance		( ) 62	( ) 64	( ) 64	( ) 57	( ) 39	( ) 50
COST				( ) 52	( ) 53	( ) 46	( ) 37	( ) 26
PUBLIC HEALTH					( ) 55	( ) 48	( ) 39	( ) 40
WATER QUALITY						( ) 48	( ) 39	( ) 41
ENVIRONMENTAL HEALTH							( ) 32	( ) 34
COST								( ) 24
SOCIAL & CULTURAL ISSUES								

Legend: Strength of relationship	
3	strong positive
2	moderate positve
1	weak positive
0	no relationship
-1	weak negative
-2	moderate negative
-3	strong negative

Figure 5. Matrix Presented to Workshop 2 for Determining the Cross Impact of the Risks and Benefits associated with the Implementation of the Current WA Policy in the Future

The participants should then be introduced to the cross impact matrices and their task for the day. The groups will be required to come to a consensus on the *strength of relationship* between each pair of risk and benefit categories in the matrix for each policy option. Using the *strength scale* outlined in the legend adjacent to the matrix, and with reference to the categories' components and their importance ratings, the groups discuss what they believe the *strength of relationship* to be. This can range from -3 being *strong negative relationship*, through 0 being *no relationship,* to +3 being *strong positive relationship.* Participants should also be given the opportunity to provide their individual ratings and these can later be compared with the group ratings[13]. This assists in keeping the groups moving along and not becoming bogged down in unproductive argument.

Participants should be reminded that their assessments are still to be made in the circumstances of the future scenario.

Group Assessment of Policy Option 1 (1 hour 30 minutes)

The time allowance here will vary according to the number of paired categories to be assessed. In the WA Case Study, the current policy had the least number of categories, and the alternative policies had increasing numbers of categories. However, it was the first policy that took the longest to complete, as this was when the different "arguments" were introduced and expressed. It was also when the participants were developing their "processes of thinking" about how to determine the *strength of relationships*.

When the groups have completed their matrices, the facilitator should combine the findings of the two groups (i.e. the mean strength ratings) and complete one feedback matrix for discussion at the end of the day. The groups' mean strength ratings should only be used if they are the same or have only minor variation. However, if there are major differences (i.e. positive and negative relationships) these should be kept separate for later discussion.

Group Assessment of Policy Options 2 and 3 (1 hour 30 minutes)

Again, the time allowance is notional and will depend on the size of the tasks to be completed. Feedback matrices should also be prepared here as described above.

Plenary discussion of Assessments (1 hour 30 minutes)

This allows 30 minutes per policy for discussion in case there are major differences between groups to resolve. If the groups provided opposing strength ratings for any pair of risk and benefit categories, these should be discussed in the plenary session to come to an agreed view[14].

The *key impact factors* that become evident as a result of the multiplication of the *strength of relationships* and the combined importances should be highlighted to the participants (e.g. see Figures 6 to 8) and invite comment about the implications of these combined categories in the implementation of the policy options.

---

[13] Generally, only minor differences were found in the WA Case Study between individual and group ratings. However, if major differences occur, this should be noted in the analysis. If the major differences are consistently associated with a particular view point or discipline, this will need to be addressed in the planning stages.

[14] If an agreed view cannot be reached, the researchers should later use a mean of all the ***individual*** strength ratings and note the key issues of contention that will need to be addressed in the planning stages.

5.1.4 Interpreting the Cross Impact Matrices

The cross impact matrices are a simple method to identify the significant issues that will need to be addressed when planning for the implementation or introduction of a policy in the future. Each category is considered in paired comparison mode with each other category in terms of its *combined importance* and *strength of relationship*. The relationship between each category is presumed to be linear. Figures 6 to 8 provide the WA Case Study example to assist in reading the following points that explain how to interpret the cross impact matrices. ***Consider the yellow cells in the matrices and the row of totals***.

- Cell inspection allows easy identification of the ***key category relationships*** that need to be considered when planning for the implementation of the policy. See the shaded cells that are identified as *Key Impact Factors* in Figures 6 to 8.
- In addition, the overall ***relative impact of each category*** can be estimated by the sum of all of its paired comparison values (i.e. see the blue TOTALS rows). It should be noted that negative numbers are added as positive numbers, as it is the *size of the relationships* that is important. The positive and negative signs indicate the *direction of relationships*.
- Prioritisation for planning can be assessed by identifying the required policy outcomes *and the key categories that will need to be addressed to ensure those outcomes*. By examining *the relationships* between categories, it can be seen what would happen if a category was addressed or ignored. For example, if there is a strong positive relationship between two categories, addressing the issues associated with one category will assist in positive results in the other category. Alternatively, if one category is ignored, it will exacerbate any negative issues in the other or reduce the effectiveness of the positive outcomes.

  The matrix should be seen as a "road map" to assist in reaching the required destination (required policy outcomes).
- Once the *categories for action* have been identified, the important individual components that were identified in the MAU analysis can contribute to specific *action plans*.

(strength of relationship) combined importance

	LOST OPPORTUNITIES	IMPLEMENTATION AND POLICING	COST	PUBLIC HEALTH	WATER QUALITY	ENVIRONMENTAL HEALTH	COST	SOCIAL & CULTURAL ISSUES
LOST OPPORTUNITIES		(3) 75	(-1) 63	(1) 65	(2) 66	(2) 58	(2) 49	(2) 40
IMPLEMENTATION AND POLICING	225		(2) 62	(2) 64	(2) 64	(2) 57	(2) 39	(2) 50
COST	-63	124		(2) 52	(2) 53	(2) 46	(2) 37	(1) 26
PUBLIC HEALTH	65	128	104		(3) 55	(0) 48	(1) 39	(1) 40
WATER QUALITY	132	128	106	165		(2) 48	(1) 39	(2) 41
ENVIRONMENTAL HEALTH	116	114	92	0	96		(0) 32	(1) 34
COST	98	78	74	39	39	0		(1) 24
SOCIAL & CULTURAL ISSUES	80	100	26	40	82	34	24	
TOTALS	779	897	589	541	748	452	352	386

Key Impact Factor (strength of relationship x combined importance)

Figure 6. Cross Impact Matrix of the Risks and Benefits associated with the Implementation of the Current WA Policy in the Future

(strength of relationship) combined importance

	PUBLIC HEALTH	WATER QUALITY	ENVIRONMENTAL HEALTH	COST	MANAGEMENT	SOCIAL ISSUES	PUBLIC HEALTH	OPPORTUNITY	MANAGEMENT	SOCIAL ISSUES	EDUCATION & RESEARCH
PUBLIC HEALTH		(3) 57	(2) 47	(2) 43	(3) 47	(0) 48	(2) 53	(2) 50	(3) 44	(2) 52	(-1) 42
WATER QUALITY	171		(2) 47	(2) 43	(2) 47	(-3) 48	(1) 54	(2) 50	(2) 45	(-3) 52	(-1) 42
ENVIRONMENTAL HEALTH	94	94		(-1) 33	(2) 38	(0) 38	(0) 44	(3) 40	(2) 35	(0) 43	(2) 33
COST	86	86	-33		(3) 33	(1) 34	(2) 39	(2) 36	(3) 30	(3) 38	(1) 28
MANAGEMENT	141	94	76	99		(2) 38	(3) 44	(3) 40	(3) 35	(3) 42	(1) 33
SOCIAL ISSUES	0	-144	0	34	76		(1) 45	(2) 41	(2) 36	(2) 43	(-1) 33
PUBLIC HEALTH	106	54	0	78	132	45		(2) 47	(2) 41	(-3) 49	(3) 39
OPPORTUNITY	100	50	120	72	120	82	74		(3) 38	(3) 45	(2) 35
MANAGEMENT	132	90	70	90	105	72	82	114		(3) 40	(3) 30
SOCIAL ISSUES	104	-156	0	114	126	86	-147	135	120		(2) 37
EDUCATION & RESEARCH	-42	-42	-66	28	33	-33	117	70	90	74	
TOTALS	976	981	553	720	1002	572	835	937	965	1078	595

Key Impact Factor (strength of relationship x combined importance)

Figure 7. Cross Impact Matrix of the Risks and Benefits associated with the Implementation of the Policy that allows Passive Recreation in the 2km RPZ in WA in the Future

(strength of relationship) combined importance

	PUBLIC HEALTH	WATER QUALITY	ENVIRONMENTAL HEALTH	MANAGEMENT	LOST OPPORTUNITIES	COST	SOCIAL ISSUES	PUBLIC HEALTH	OPPORTUNITY	MANAGEMENT	SOCIAL ISSUES	EDUCATION & RESEARCH
PUBLIC HEALTH		(3) 31	(2) 30	(-3) 30	(-3) 27	(3) 31	(2) 23	(3) 34	(-3) 39	(0) 40	(2) 29	(-1) 31
WATER QUALITY	93		(2) 33	(-3) 32	(-2) 29	(3) 33	(2) 25	(3) 36	(3) 41	(-1) 42	(2) 31	(-1) 33
ENVIRONMENTAL HEALTH	60	66		(3) 32	(-2) 28	(3) 32	(1) 25	(0) 35	(3) 41	(-1) 41	(2) 30	(-1) 33
MANAGEMENT	-90	-96	96		(-2) 28	(3) 32	(1) 24	(-3) 35	(3) 40	(2) 41	(3) 30	(-1) 33
LOST OPPORTUNITIES	-81	-58	-56	-56		(0) 29	(1) 21	(2) 31	(3) 37	(1) 38	(1) 26	(0) 29
COST	93	99	96	96	0		(2) 25	(-3) 35	(3) 41	(3) 42	(3) 30	(2) 33
SOCIAL ISSUES	46	50	25	24	21	50		(0) 28	(3) 33	(2) 34	(3) 23	(-1) 26
PUBLIC HEALTH	102	108	0	-105	62	-105	0		(-3) 44	(2) 44	(2) 33	(3) 36
OPPORTUNITY	-117	123	123	120	111	123	99	-132		(3) 50	(2) 39	(1) 42
MANAGEMENT	0	-42	-41	82	38	126	68	88	150		(2) 39	(2) 42
SOCIAL ISSUES	58	62	60	90	26	90	69	66	78	78		(1) 31
EDUCATION & RESEARCH	-31	-33	-33	-33	0	66	-26	36	42	84	31	
TOTALS	771	830	656	888	509	944	478	789	1218	797	708	415

Key Impact Factor (strength of relationship x combined importance)

Figure 8. Cross Impact Matrix of the Risks and Benefits Associated with the Implementation of the Policy that allows Off-Road Access in the Outer Catchment in WA in the Future

5.1.5 Cross Impact Analysis of the Policies in the WA Case Study

5.1.5.1 Current Policy

From Figure 6 and the examination of totals and significant relationships, it is apparent that the risk category, *Implementation and Policing,* is the single largest influence, by a substantial margin, on all risk-benefit categories of the current policy under the future scenario. It is particularly manifest in its interaction with the risk category *Lost Opportunities.* The benefit category, *Water Quality,* is also highly influential, particularly in its relationship with the benefit category *Public Health.*

Water Quality and *Public Health* are the major reasons underpinning the arguments for the justification of the current policy, and public support of the policy is based on these outcomes. From the matrix, it is apparent, that under the future scenario, their relationships with *Implementation and Policing* and *Lost Opportunities* are important. The matrix also suggests that there will be *Cost* implications associated with both *Implementation and Policing* and *Lost Opportunities.* Examination of the components of each category show why these relationships in the future are important.

IMPLEMENTATION AND POLICING

- False sense of security of drinking water safety
- Overload on outer catchment facilities
- Inability to police and enforce the policy
- It's "fun to disobey"
- Public dismissal of the policy

COST

- High cost of policing the policy
- Legal risks associated with prohibited access
- Loss of near-water land for other uses
- Inefficient use of land for water supply compared with manufactured sources
- Opportunity cost for alternative land uses.

LOST OPPORTUNITIES

- General recreation demand not met
- Nature based recreation not met
- Eco-tourism opportunities lost
- Reduced enjoyment through overload of insufficient facilities
- Lost opportunity to educate the public about catchments
- Lost opportunity for eco-education activities

WATER QUALITY

- High quality water with little treatment
- Lack of reliance on advanced water treatment
- Avoids possible contaminants (prevention better than the cure)
- Reduced fire risks
- 2km RPZ filters the water
- Reduced access to feeder streams with the 2km RPZ

PUBLIC HEALTH

- Low human health risks
- Human health benefits from recreation available outside the 2km RPZ
- Provides source diversity to ensure supply under conditions of uncertainty

The pressure resulting from an inadequately managed catchment due to increasing issues associated with the risk components of *Implementation and Policing* will rise over the next 30 years. Ongoing concern by increasing population numbers over *Lost Opportunities* will put further pressure on *Implementation and Policing*. This will be to the potential detriment of both *Water Quality* and *Public Health*. This indicates that the current policy and its desired outcomes are seen to be difficult to sustain in the long term.

Looking at innovative and sustainable modes of *Implementation and Policing* should be a priority if this policy is to be maintained in the future. Ensuring there are alternative locations and facilities for people to enjoy the recreational *Opportunities* that are not available at drinking water catchments will also be important.

5.1.5.2 Policy that allows Passive recreation in the 2km RPZ

It is clear that when passive recreation is considered for the 2 km RPZ, the range of important interactions creates wider challenges than for the current policy. This is also demonstrated by the greater number of categories that were identified for this alternative policy.

From Figure 7 and the examination of totals and key impact relationships, it is clear that *Social Issues* (benefit), and *Management* (risk and benefit) and their relationships with *Water Quality* (risk) and *Public Health* (risk and benefit) are the major features in the assessment of the 2km RPZ access policy. However, it is apparent that *Social Issues* dominates this assessment.

The arguments for not implementing this policy are clearly displayed in the component risks of *Public Health* and *Water Quality* and a strong relationship between these two categories exists in the matrix. There are also strong relationships between *Management*, both risk and benefit, and *Public Health*. Related to all these is *Social Issues* which is also strongly related to *Opportunity*. The components of these categories shown below assist in understanding the relationships.

SOCIAL ISSUES

- Provides wilderness and nature based experiences
- Conduit for social networking
- Increased visual amenity close to water
- Builds community support for natural experiences

OPPORTUNITY

- Recognises the overall value of the catchment resource
- Meets greater recreational demand
- Increased eco-tourism opportunities
- Greater accountability of actions between recreators (e.g. community "policing")
- Potential for development of improved water treatment technologies

MANAGEMENT

- Risk of "prohibited activity creep effect" (i.e. prohibited accessing of water from the RPZ or motorised activities into RPZ)
- Risk of "policy creep effect" (i.e. once the policy is relaxed, it's easy to relax it further with greater recreational access and alternative land uses)
- Increased complexity of management
- Increased complacency about water quality security
- Harder to police
- Increased water quality monitoring

MANAGEMENT

- Will require better management
- Will require best practice facilities to reduce risks
- Potential to introduce fees to generate revenue for improved management and facilities
- Increased requirement for cross-agency collaboration

WATER QUALITY

- Increased human contamination
- Need for higher level treatment
- Lack of ability for 2km RPZ to filter the water
- Increased likelihood of human contact with water
- Increased risk of fire

PUBLIC HEALTH

- Increased management of health risks

PUBLIC HEALTH

- Health benefits from increased recreational opportunities

The future scenario indicates that the demands of *Social Issues* and providing *Opportunities* are likely to become more pressing and will threaten the management of the RPZ. Meeting both the recreation and social needs while maintaining publicly acceptable *Water Quality* could be difficult. Also, providing greater access to the catchment will increase the number of recreators and likely interfere with the wilderness and nature based experiences being sought. *Management* of this situation will provide its challenges as shown in the risk components, but *Management* can also be seen as having positive outcomes as shown in the benefit components.

If a sustainable balance is to be achieved in a future of growing social pressures, recreation, tourism, health and water professionals will need to work closely to address the key relationships shown in the matrix. This will need to balance recreational demand and experience in the region with prudent catchment management to protect water quality and public health.

5.1.5.3 Policy that allows Off-Road Access in the Outer Catchment

From Figure 8, the benefit *Opportunity* dominates the assessment of the policy that allows off-road access to the outer catchment. *Cost* (risk) and *Management* (both risk and benefit) also have important relationships. *Opportunity* is strongly related to *Environmental Health* (risk) and has negative relationships with *Public Health* (both risk and benefit). While *Opportunity* is a benefit, it has strong relationships with a number of important risk categories. Again *Water Quality* features in the assessment.

Examination of the components of these categories explain these key relationships.

OPPORTUNITY

- Recreational access for off-road vehicles
- Potential to create long distance off-road vehicle tracks
- Greater accountability of actions between recreators (e.g. community "policing")
- Potential for development of improved water treatment technologies

MANAGEMENT

- Greater control (e.g. through zoning of motorised recreation)
- Will require increased access to facilities (e.g. ablutions)
- Potential to introduce fees to generate revenue for improved management and facilities
- Recognition of what is already happening

MANAGEMENT

- Increased complexity of management
- Increased complexity of policing (i.e. registered and non-registered vehicles)
- Difficulty in meeting created demand

COSTS

- Increased treatment costs
- Increased management costs
- Increased policing costs
- Increased road and track maintenance costs
- Increased rehabilitation costs
- Inefficient use of land for water supply compared with manufactured sources
- Opportunity cost for alternative land uses

ENVIRONMENTAL HEALTH

- Increased spread of weeds
- Increased environmental degradation

WATER QUALITY

- Increased contamination from motorised vehicles (i.e. hydrocarbons etc.)
- Increased risk of fire
- Increased erosion
- Need for increased treatment

PUBLIC HEALTH

- Increased management of health risks
- Increased safety risks associated with motorised vehicles (i.e. to both passive and active recreators)

PUBLIC HEALTH

- Health benefits from increased recreational opportunities

If this policy were to be adopted there would be a clear need for a much greater effort to be paid to planning and definition, both in regard to facilities for off-road vehicles, and the prevention of conflict between and within motorised and non-motorised recreators. However, the matrix indicates that the major decision is whether the benefits attained from the additional recreation opportunities and the positive management and public health outcomes could be reconciled in the face of water quality and public health risks, management challenges, environmental degradation and all the associated costs.

6. REFERENCES

Ackerman F and Heinzerling L (2004) *Priceless: On Knowing the Price of Everything and the Value of Nothing*. New Press: New York.

Ananda J and Herath G (2005) Evaluating public risk preferences in forest land-use choices using multiattribute utility theory. *Ecological Economics*, 55, 408-419.

Australia 21 (2008) *A National Ecosystem Services Strategy for Australia*. www.australia21.org.au

Bamberg S and Moser G (2007) Twenty-years after Hines, Hungerford and Tamera: A new meta-analysis of psycho-social determinants of pro-environmental behaviour. *Journal of Environmental Psychology*. 27, 14-25.

Bammer G and Smithson M (Eds.) (2008) *Uncertainty and Risk: Multidisciplinary Perspectives*. Earthscan: London.

Bateman IJ, Carson RT, Day B, Haneman WM Hanley N, Hett T et al. (2002) *Economic Valuation with Stated Preference Techniques: A Manual.* Edward Elgar: Cheltenham.

Bateman IJ, Cole MA, Georgiou S, and Hadley DJ (2006) Comparing contingent valuation and contingent ranking: A case study considering the benefits of urban river water quality improvements. *Journal of Environmental Management,* 79, 221-231.

Bestard AB and Font AR (2009) Environmental diversity in recreational choice modelling. *Ecological Economics*, 68, 2743-2750.

Biloslavo R and Dolinsek S (2010) Scenario planning for climate strategies development by integrating group Delphi, AHP and dynamic fuzzy cognitive ma.pos. *Foresight,* 12, 38-48.

Bruce D (2006a) *South West of Western Australia Water Catchment Management Issues, Drinking Water Policy and Logue Brook Dam: Community Survey Results.* Report prepared for the Department of Water, WA.

Bruce D (2006b) *South West of Western Australia Water Catchment Management Issues, Drinking Water Policy and Logue Brook Dam: Dialogue Forum Deliberative Survey Results.* Report prepared for the Department of Water, WA.

Brunsen MW and Shelby B (1993) Recreation substitutability: A research agenda. *Leisure Sciences.* 15, 67-74.

Burger J, Gochfeld M, Pletnikoff K, Snigaroff R, Snigaroff D and Stamm T (2008) Ecocultural attributes: Evaluating ecological degradation in terms of ecological goods and services versus subsistence and tribal values. *Risk Analysis*, 28 (5), 1261-1271.

Castle D (2000) Limitations on an inclusive definition of ecosystem-human health. *Environmental Ethics*, 5(2), 153-161.

Cilimburg A, Monz C And Kehoe S (2000) Wildland recreation and human waste: A review of problems, practices, concerns. *Environmental Management*, 25 (6), 587-598.

Connerley CE (1986) Growth management concern: The impact of its definition on support for local growth controls. *Environment and Behavior*, 18 (6), 707-732.

Damodar S (1995) *Psychological Aspects of the Polarisation Phenomenon,* Concept: New Delhi.

Dawes RM and Corrigan B (1974) Linear models in decision making. *Psychological Bulletin*, 81, 95-106.

Dawes RM (1979) The robust beauty of improper linear models in decision making, *American Psychologist,* 34, 571-582.

Dietz T, Dan A, and Shwom R (2007) Support for climate change policy: Social psychological and social structural differences. *Rural Sociology,* 72 (2), 185-214.

DSR (2006) (WA Department of Sport and Recreation) Strategic Plan 2006-2010. http://dsr.wa.gov.au/

Eaton E and Syme GJ (1989) *Murray Shire Community Development Survey.* Report to the WA Department of Resources Development. CSIRO: Perth.

Edwards W (1979) Multi-attribute utility measurement: Evaluating desegregation Plans in a Highly Political Context. In R. Perloff (Ed.) *Evaluator interventions: Pros and Cons,* (pp 13-53) Sage: Beverly Hills.

Edwards W and von Winterfeldt D (1987) Public values in risk debates. *Risk Analysis*, 7 (2), 141-158.

Egoh B, Douget M, Reyers B, Knight AT, Cowling RM, van Joersveld AS and Wetz A (2007) Integrating ecosystem services into conservation assessments: A review. *Ecological Economics*, 63, 714-721.

Eisenhauer BW, Krannich RS and Blahna DJ (2000) Attachments to special places on public lands: An analysis of activities, reason for attachments, and community connections. *Society and Natural Resources,* 13, 421-441.

Enzer S (1984) Anticipating the unpredictable. *Technological Forecasting and Social Change*, 26, 201-205.

Enzer S and Alter S (1978) Cross Impact analysis and classical probability: The question of consistency. *Futures*, 10, 227-239.

Ferguson CM, Croke BFW, Beatson PJ, Ashbolt NJ and Deere DA (2007) Development of a process based model to predict pathogen budgets for the Sydney drinking water catchment. *Journal of Water and Health,* 5 (2), 187-208.

Ferguson C, de Roda Housman AM, Altavilla N, Deere D and Ashbolt N (2003) Fate and transport of surface water pathogens in watersheds. *Critical Reviews in Environmental Science and Technology,* 33 (3), 299-361

Fink CL (1989) A spatial model of the mere exposure effect. *Communication Research*, 16 (6), 746-769.

Finucane M, Alhakami A, Slovic P and Johnson S (2000) The affect heuristic in judgments of risks and benefits. *Journal of Behavioural Decision Making*, 13, 1-17.

Fisher B, Turner RV and Morling P (2009) Defining and classifying ecosystem services. *Ecological Economics*, 68, 643-653.

Foster V and Mourato S (2000) Valuing the multiple impacts of pesticide use in the UK: A contingent ranking approach. *Journal of Agricultural Economics*, 49, 1-15.

Fu C and Yang S-L (2010) The group consensus based evidential reasoning approach for multiple-attribute group decision analysis. *European Journal of Operational Research*, 206, 601-608.

Gardner PC and Edwards W (1975) Public Values: Multi-attribute utility measurement for social decision making. In MF Kaplan and S Schwartz (Eds) *Human Judgment and Decision Processes*. (pp 1-36). Academic Press: New York.

Groves RM (2006) Nonresponse rates and nonresponse bias in household surveys. *Public Opinion Quarterly,* 70 (5), 646-675.

Hajkowicz S and Collins K (2007) A review of multiple criteria analysis for water resources planning and management. *Water Resources Management*, 21 (9), 1573-1650.

Hogarth RM and Natalia K (2007) Heuristic and linear models of judgment: Matching rules and environments. *Psychological Review*, 114, 753-758.

Hughes M, Zulfa M and Carlsen J (2008) *A review of recreation in public drinking water catchment areas in the Southwest Region of Western Australia.* Curtin Sustainable Tourism Centre, Curtin University, Perth.

Ivanova G and Tranter B (2008) Paying for environmental protection in a cross-national perspective. *Australian Journal of Political Science*, 43 (2), 169-188.

Ivey JL, de Loe R, Kreutzwiser R And Ferreyra C (2006) An institutional perspective on local capacity for source water protection. *Geoforum*, 37, 944-957.

Johnson BB (2008) Public views on drinking water quality standards. *Risk Analysis*, 28 (6), 1515-1530.

Jorgensen BS and Stedman RC (2006) A comparative analysis of predictors of sense of place dimensions: Attachment to, dependence on, and identification with lakeside properties. *Journal of Environmental Management,* 79, 316-327.

Jorgensen BS and Syme GJ (2000) Protest responses and willingness to pay: attitude towards paying for stormwater pollution abatement. *Ecological Economics,* 33, 251-265.

Kahneman D And Tversky A (1979) Prospect theory: An analysis of decision making under uncertainty. *Econometrica* , 47, 263-291.

Kahneman D, Ritov I, Schkade D (1999) Economic preferences or attitude expressions? An analysis of dollar responses to public issues. *Journal of Risk and Uncertainty,* 19, 203-235.

Karpiak CP and Baril GL (2008) Moral reasoning and concern for the environment. *Journal of Environmental Psychology,* 28, 203-208.

Kline RB (2005) *Principles and practice of Structural Equation Modelling.* (2nd Ed.), Guilford, The Guilford Press: New York.

Kroeger T and Casey F (2007) An assessment of market-based approaches to providing ecosystem services on agricultural lands. *Ecological Economics,* 64, 321-332.

Krosnick JA (1988) The role of attitude importance in social evaluation: A study of policy preferences, presidential candidate evaluations and voting behavior. *Journal of Personality and Social Psychology*, 55, 196-210.

Lawrence RL, David SE, and Stankey GH (1997) Procedural justice and public involvement in natural resources decision making. *Society and Natural Resources*, 10, 577-589.

Leder K, Sinclair MI and McNeil JJ (2002) Water and the environment: a natural resource or a limited luxury? *Medical Journal of Australia*, 177, 609-613.

Loomis J (2000) Environmental valuation techniques in water resources management. *Journal of Water Resources Planning and Management*, 126(6), 339-344.

Lumley S and Hercock M (2000) Locational and socio-economic variation in public perceptions of economics and environment. *GeoJournal,* 51, 235-244.

Mallawaarachchi T, Balmey RK, Morrison MD, Johnson AKL and Bennett JW (2001) Community values for environmental protection in Northern Australia: A choice modelling study. Journal of *Environmental Management*, 62, 301-316.

Manfredo MJ and Anderson D (1987) The influence of activity importance and similarity on perception of recreational substitutes. *Leisure Sciences,* 9, 77-86.

McIntyre W, Tucker D, Green M, Syme G, Bates L, Porter N, and Nancarrow B (2006) *Water Benefits Accounting and Assessment: Lake Mulwala Case Study.* CSIRO Water for a Healthy Country National Research Flagship, Land and Water: Perth.

Millennium Assessment, (2005) *Ecosystems and Human Wellbeing Synthesis.* World Resources Institute, Washington, DC.

Miller R, Bennett B, Birrell J, and Deere D (2006) *Recreational Access to Drinking Water Catchments and Storages in Australia.* CRC Water Quality and Treatment. Research Report No. 24.

Mitchell RC and Carson RT (1989) *Using Surveys to Value Public Goods: The Contingent Valuation Method.* Resources for the Future: Washington.

Moran C, Syme G, Hatfield-Dodds S, Porter N, Kington E and Bates L (2004) *On Defining and Measuring the Benefits from Water.* 2nd IWA Leading-Edge Conference on Sustainability: Sustainability in Water Limited Environments – LES 2004. 8-10 November 2004. Sydney: Australia.

Morrison MD, Blamey RK and Bennett JW (2000) Minimising payment vehicle bias in contingent valuation studies. *Environmental and Resource Economics*, 16, 407-422.

Nancarrow BE, Leviston Z And Tucker DI (2009) Measuring the predictors of communities' behavioural decisions for potable reuse of wastewater. *Water Science and Technology,* 60 (12), 3199-3209.

Petty RE and Krosnik JA (Eds) (1995) *Attitude Strength.* Lawrence Erlbaum, New Jersey.

Pollard SJT, Strutt JE, MacGillivray BH, Hamilton PD and Hrudey SE (2004) Risk analysis and management: A review of drivers, tools and techniques. *Process Safety and Environmental Protection*, 82 (6), 453-462.

Postel SL and Thompson BH (2005) Watershed protection: Capturing the benefits of nature's water supply services. *Natural Resources Forum*, 29, 98-108.

Proctor W and Drechsler M (2006) Deliberative multicriteria evaluation. *Environment and planning: Government and Policy,* 24(2), 169-190.

Randhir T and Shriver DM (2009) Deliberative valuation without prices: A multiattribute prioritisation for watershed ecosystem management. *Ecological Economics*, 68, 3042-3051.

Randhir TD, O'Connor R, Penner PR, and Goodwin DW (2001) A watershed-based land prioritization model for water supply protection. *Forest Ecology and Management*, 143, 47-56.

Rolfe J and Bennett J (Eds) (2006) *Choice modelling and the transfer of environmental values.* Edwin Elgar: Cheltenham.

Ross V (2005) *The Determinants of Trust and Satisfaction with Drinking Water Quality.* ARCWIS Occasional Paper, CSIRO Land and Water: Perth.

Rothstein E and Kiyosaki D (2003) Development of a strategic plan: portfolio management for public utilities. *Journal of the American Wastewater Association*, 95 (1), 52-65.

Saaty T (1999) *Decision Making for Leaders* (3rd Ed.). RWS Publications: Pittsburgh,

Schultz MT (2001) A critique of the EPA's index of watershed indicators. *Journal of Environmental Management,* 62, 429-442.

Sherif M (1961) *Social judgment: Assimilation and contrast effects in communication and attitude change.* Yale University Press: New Haven.

Sorice MG, Oh C-O, and Ditton RB (2009) Exploring level of support for management restrictions using a self-classification measure of recreation specialization. *Leisure Sciences,* 31, 107-123.

Spash CJ, Urama K, Burton R, Kenyon W, Shannon P and Hill G (2009) Motives behind willingness to pay for improving biodiversity to a water ecosystem: Economics, ethics and social psychology. *Ecological Economics,* 68, 953-964.

Stebbins RA (1992) *Amateurs, Professionals and Serious Leisure*. Queen's University Press: Montreal.

Steinberg PE and Clarke G (1999) Troubled water? Acquiescence, conflict and the politics of place in Watershed management. *Political Geography*, 18 (4) 477-508.

Sullivan R (1998) Assessing the acceptability of environmental risk – A public policy perspective. *Australian Journal of Environmental Management.* 5 (2), 72-80.

Svensson M (2009) Precautionary behaviour and willingness to pay for a mortality risk reduction: Searching for the expected relationship. *Journal of Risk and Uncertainty*, 39, 65-85.

Syme GJ (1995) Community acceptance of risk: trust, liability and consent. In P.Heinrichs and R. Fell (Eds.) *Acceptable Risk for Major Infrastructure*. A.A. Balkema, Rotterdam.

Syme GJ and Bishop BJ (1992) *Community Perceptions of Dam Safety Issues: A Preliminary Study.* CSIRO Division of Water Resources. DWR Consultancy Report No. 92/32.

Syme GJ and Nancarrow BE (2005) Sharing the pain of water allocation: Creating consent by taking fairness and justice seriously. In D. Coyle and J. Hillier (Eds) *Consent and Consensus: Politics, Media and Governance in Twentieth Century Australia.* (pp 371-387), API Network Curtin University of Technology: Perth.

Syme GJ and Nancarrow BE (2008) Justice and allocation of benefits from water. *Social Alternatives,* 27 (3), 21-25.

Syme GJ and Nancarrow BE (2009) *Climate Change In WA: Longitudinal Study of Community Attitudes to and Preferences for Mitigation and Adaptation Policies – The Baseline Measure*. Report to the Department of Environment and Conservation, WA. Edith Cowan University: Perth.

Syme GJ, Nancarrow BE and McCreddin JA (1999) Defining the Components of Fairness in the Allocation of Water to Environmental and Human Uses. *Journal of Environmental Management*, 57, 51-70.

Tanner RJ, Friemund WA, Borrie WT and Moisey RN (2008) A meta-study of the values of visitors to four protected areas in the western United States. *Leisure Sciences*, 30, 377-390.

Teisel MF and O'Brien K (2003) Who cares and who acts?: Outdoor recreationists exhibit different levels of environmental concern and behavior. *Environment and Behavior*, *35*, 506-522.

Tsaur S-H and Liang Y-W (2008) Serious leisure and recreation specialisation. *Leisure Sciences*, 30, 325-341.

Tucker D, Johnston C, Leviston Z, Jorgensen BS and Nancarrow BE (2006) *Sense of Place: Towards a Methodology to Value Externalities Associated with Urban Water Systems.* CSIRO Water for a Healthy Country National Research Flagship, Land and Water: Perth.

Tyler TR and Blader SL (2000) Cooperation in Groups: Procedural Justice, Social Identity, and Behavioral Engagement. Psychology Press: Philadelphia

Ungar S (2000) Knowledge, ignorance and the popular culture: Climate change versus the ozone hole. *Public Understanding of Science,* 9(3), 297-312.

Weiss JW and Weiss DJ (2009) *A Science of Decision Making.* Oxford University Press: Oxford.

Weiss JW, Weiss DJ and Edwards W (2010) A descriptive multi-attribute utility model for everyday decisions. *Theory and Decision*, 68, 101-114.

Wilke, H.A.M. (1991) Greed, efficiency and fairness in resource management situations. *European Review of Social Psychology*, 2, 165-188.

Williams KD and Syme GJ (1989) *Consumer Attitudes towards Drinking Water Quality and Water Restrictions Policy.* Report to the Australian Water Resources Council, CSIRO: Perth.

Wong EY, Ponce RA, Farrow S, Bartell SM, Lee RC and Faustmann EM (2003) Comparative risk and policy analysis in environmental health. *Risk Analysis,* 23, 1337-1349.

Wright G and Goodwin P (1999) Future focussed thinking: Combining scenario planning with decision analysis. *Journal of Multi-Criteria Decision Making*, 8, 311-321.

WSAA (2008) *Effects of Recreational Activities on Source Water Protection Areas: A Literature Review*. Occasional Paper No. 22, WSAA, Melbourne.

Yearwood J and Stranieri A (2010) Group structured reasoning for coalescing group decisions. *Group Decision and Negotiation*, 19, 77-105.

APPENDIX 1
NOTES OF SCOPING INTERVIEWS WITH PROFESSIONALS IN WA

The following details the key outcomes of face to face in-depth interviews with ten professionals associated with drinking water production, regulation, catchment management (including a catchment ranger) and biodiversity.

It is presented as dot points, but organised under theme headings.

HEALTH AND TREATMENT

- Community considers that if water is under pressure and it looks OK then it is OK.
- Last health scare was in 1896 with the water supply from the Victoria reservoir. From then on, it was decided to keep everyone out of the catchment and there have been no problems since. 100 years of safe water supply because it's allowed to be a "natural system".
- Minimum treatment required – water in most of the dams already meets DWQ guidelines and only requires chlorination and no filtration – cheapest and most stable supply.
- There is no question that relying on treatment is a second best option.
- Groundwater has a good natural filtration system – by the time it is 30m from a septic system, it's OK to drink. Safest, most reliable source of water we could ask for.
- By the time water tests are done and problems found, people will already be contaminated. There are very short holding times for water in tanks and reservoirs.
- Hard to pin down where an incident comes from (e.g. swimming pools) and hence a direct link is hard to establish.
- Vehicles bogging in either the dam or the feeder streams cause turbidity. Even one vehicle causes considerable turbidity and if it reaches a certain level, chlorination cannot be guaranteed to treat the water.
- 4WD vehicles destroy the understory of vegetation and that's a major protection against pathogens reaching the water.

SYSTEM OPERATION AND RISK

- Perth has lots of dams, not just one big one like Sydney's Warragamba, so there is the ability to take dams off line if there is a quality issue or incident. For example, some dams are taken offline at certain times of the year because of contamination from natural occurrences in the catchments.
- Water Corp is a primary producer (produces water) just like a wheat farmer. Both have catchments and you can't drive a 4WD across the farmer's catchment, so why should you be able to do so in Water Corp's catchments.
- What do you do with contaminated water? It takes 2 hours to turn off a water main and there's a lot of water. Where do you put it and how long does it take to decontaminate a pipeline?
- How long does it take for a community to regain confidence in the safety of the water?
- It's all about risk minimisation. If greater treatment is required, it's with man-made membranes and systems and with the DALI approach, deaths are accepted. Is this the right approach as all sorts of trade-offs will occur that could compromise water sources in the long term (i.e. risk assessment is not the approach).
- We have no knowledge of how people will react to an incident as there is no previous experience.
- With non-point pollution, it's hard for people to comprehend risk.
- Could set a maximum contamination number for recreators.
- The idea of adding a risk and then mitigating the risk is flawed. If the mitigation fails, the risk is still there.
- There is Risk Management or Preventative Risk Management – there's a difference and the Australian DWQ Guidelines recommend the second.

- Don't put the risk there in the first place if it's unnecessary – there are other places for people. An ounce of prevention is better than a pound of cure.
- Recreation is the thin edge of the wedge. Once treatment is added, it leaves it open for other uses such as industry and prisons. Once one risk goes in, others can be added.
- There are also unknown risks.
- It's not the steady state conditions that are the problem. It's the storm events etc. that have to be planned for.
- Our catchments can't cope with potential contamination from body contact with water.
- There is a lack of understanding of micro-biological contamination and camping is a potential source of this. The catchments are massive, but even contamination from high in the catchment can reach the water. However, people like to camp by water's edge and don't understand the risk. So it's not just a matter of walking in the *Reservoir Protection Zone* (RPZ), there is a likelihood that people will want to camp there too.
- At Victoria Dam, rangers found 10kg of meat in the water last week, turkey legs and babies nappies. This is what happens when people get close to the reservoirs. It's a huge risk.
- Last week there was a 4WD vehicle bogged at the back of Victoria Dam. If the sump had broken the oil would have polluted the whole dam and with the potential of getting into the mains. There would be big problems getting it out of the mains.

PLANNING

- Perth has been designed as a 20km wide strip up and down the coast which provides a legacy greenfield area (catchments and forests) only 30 minutes from the city. Within that area there is only a 2km RPZ around each dam where people are not allowed to go.
- Why 2km chosen for RPZ? You can't see the dam from 2kms away and so you are less likely to want to walk towards it. Also if you can't see the dam, it's assumed that contaminants are less likely to reach the dam. Perhaps the RPZ would be better arranged through land contours and tributaries.
- Unsure of the justification of the 2km RPZ but it is a good buffer to keep people away and a good buffer for erosion activities. Obviously pathogens can still come down the feeder streams. But if the zone were 200 meters, people would see the water and want to go through to it.
- Need evidence and an argument for why the RPZ is 2 km. It must also depend on topography and land use.
- What is the justification for the RPZ boundary. Maybe different zones at different distances could support different activities e.g. 200 metres for bushwalking, and incrementally further from the water according to passive or active.
- There are opportunities to restrict or enhance recreation over groundwater, depending on the vulnerability.
- Classify catchments according to vulnerability and manage accordingly.
- Priorities are clear: 1- DWQ; 2 – ecosystem protection; 3 – recreation. The recreation allowed in 3 must not compromise priorities 1 & 2.
- Need to see what recreation can occur outside the catchments first, then think about inside the catchments.
- Where can people go and not cause a problem, and what activities are compatible?
- Make things more attractive - assign certain dams to certain activities that are compatible and reduce the conflict between different recreational users.
- Recreators can currently use the feeder streams in the outer catchment, but DSR are suggesting no water-based activities in the outer catchments – limited to passive recreation.
- Other water bodies, estuaries and rivers are not developed to their potential. Everyone likes to be next to a water body.
- Open up the Water Corporation catchments that are not drinking water catchments. Give them up incrementally, as other areas become fully utilised.

- There are some possibilities of handing over some water bodies in the SW for recreation, but in other cases where they are close to towns and the only source of water for the towns, then protection is more important. If it becomes contaminated, there is no other source of DW for the towns.
- Water Corporation could pay for upgrading facilities at appropriate water bodies to save their drinking water (DW) dams.

GOVERNANCE & MANAGEMENT

- Policy 13 is right and it meets scientific and community expectations. It co-exists with DEC Policy 18.
- The Catchment Protection Act pre-dates the CALM Act.
- There are a set of By-Laws to go with the policy. A lot are hard to enforce but the RPZ is easy to enforce.
- Catchment managers operate under two different sets of Department of Water (DoW) By-Laws: Metropolitan By-Laws and the Country Area Water Supply. The Metropolitan By-Laws clearly spells out the RPZ while the Country Area Water Supply talks about a no go zone but there is no legal "keep out" zone. They can walk to the water's edge and can camp within 300 metres of the water's edge. It's a big difference.
- Up until two years ago Mundaring Weir was under the Country Area Water Supply, and is still the case on the eastern side in the Shires of York and Northam. Policing of the RPZ started 18 months ago. There is a big difference between Mundaring and Canning which has always had an RPZ, especially around the water's edge.
- It's an inter-departmental responsibility, but some in DoW consider it is a DoH problem. Water Corp coordinates with DEC and Fisheries.
- Need to provide conditions of use e.g. leave only footprints.
- Need to think before banning types of recreation and whether it can be managed. For example, DEC banned off-road vehicles in the Logue Brook catchment whereas DoW would have been prepared to manage them. Logue Brook wasn't a win for everyone.
- DEC have strict controls in National Parks, but allow more freedom in State Forests.
- DEC guidelines say to stick to tracks and designated camp sites.
- Must camp at designated camp sites as we need to know where people are in case of evacuation for fire etc. But bushwalkers want to follow any tracks, including animal or natural. Perhaps they could do it, subject to permits, so we know they are there.
- More consistent and stronger regulations are required. Current fines are virtually nothing ($40 - $200) and people consider it is worth paying for the fun.
- When it says "No Camping", the sign should also say where the nearest campsite is.
- People want facilities at designated sites e.g. picnic sites with toilets.
- Need better management at peak times in the current locations, rather than opening up other areas.
- Harvey Dam is well managed and looks good. Other places are not maintained, insufficient facilities (e.g. boat launching facilities), toilets are smashed, camp fires are left burning, and so people will pollute.
- Hoons go to the places that are poorly kept and managed and so the families stay away. If it is attractive for families, they will go there, and the hoons will stay away.
- After dark is a problem as that's when facilities get destroyed. There needs to be tougher legislation to keep hoons away.
- The best way to control the hoons is with infringement notices but it costs a lot of money to issue infringement notices and WC won't get the money back. But it means people have to go to court and that takes a lot of time.
- We spend a lot of time talking about the small percentage that cause trouble. But in the long term it's how we manage the bigger groups and ensure DWQ. The more the bigger groups get access to the catchment, the more the smaller groups will follow.
- Dams take up only a small proportion of the catchment areas. Put better facilities in the outer catchments – more tracks, water tanks etc. and entice people away from the water bodies.

- Government needs to show they are serious about the RPZ. Fence it off, keep individuals out, and remove feral animals to properly protect the water. Catchments in Victoria are fenced and people don't question it.
- Government needs to be seen to be protecting DW for the population, then they'd get support. It's no good just talking about it.
- There is not enough funding at the moment for facilities and management. Policing is not sufficient.
- Whatever we do will take money to manage it well.
- Try to give groups one point of call for permits. They firstly apply to DEC who makes the decision and then they are sent to WC. In past times, DEC might refuse them their permits and then they'd come to the WC who didn't know they'd been refused. Groups tended to play one department off against another.
- DEC is the manager of State Forests and the majority of the catchments are State Forests. If someone wants to get access to State Forests, they have to go to DEC. WC works closely with DEC and have a combined set of conditions. Should be no discrepancy between the two.
- It can be hard –e.g. there are Rave Parties in the catchment that DEC try to manage and suggest they could be held in non-designated camp sites. However WC has to say no as it's not a designated camp site. These events have the potential to suddenly grow and there are not enough toilets etc. Also, this would set a precedence for others.
- Currently WC is not a recreation manager and so resources are sufficient for interactions with other departments. But if WC were to become more involved in recreation management then more resources would be needed.
- DSR and the potable water people now agree so it's the recreators against everyone.
- Each dam needs a management plan that clearly says what you can and can't do and why. They should be developed across Government Departments and with the community. Rules need to be different for each dam and there needs to be a consistent Government approach and not conflicting information according to different Acts and Legislation for each Department.

POLICING

- Illegal activities in the prohibited zones are increasing. Rangers recently worked on a Saturday and Sunday night and got 8 prosecutions with By-Law breaches. Some of what they were doing was dreadful, and very close to the treatment plant.
- Rangers work at all times – not advertised but spot checks day or night, any day. Rangers live on certain dam sites.
- There is a lot of illegal marroning so policing needs to be after hours and WC works closely with Fisheries. All manner of polluting materials are used in the water to attract marron (e.g. blood and bone, chicken pellets, cat and dog food, kangaroo, sheep, turkey etc.). A lot of prosecutions each year, but it's hard to catch them.
- Man is a poacher! He's been poaching for a thousand years.
- Have to be careful not to set a precedence. If you allow one person or one family to do something, then many others will want to do it too.
- There are large, clear, reflective signs on 99.9% of the tracks not to go further into the prohibited RPZ. People can't miss them. If they are coming through the bush, they probably won't know they entered the prohibited zone but most people use the tracks.
- A lot of people rely on GPS and the RPZs are not shown, but we are working on this.
- 278 prosecutions last year under DoW By-Laws and there are also written and verbal warnings. The prosecutions result in a $100 fine plus $226 court costs. No-one under 18 is prosecuted.
- If there are a few prosecutions in a short period of time, word of mouth occurs and things quieten down for a while.
- Not too many repeat offenders in marroning – and not many repeat offenders in other areas.

- Rangers believe they have it under control. If they had double the people, it wouldn't result in much greater improvement. Fatigue through driving long distances is an issue and they are looking at ways to reduce driving. There are new technologies coming to hand which can show when someone goes down a track. This isn't too far off and will help policing.

RECREATION ISSUES

- Bushwalkers aren't "hoons" with their cartons of tinnies. They are into health. Is there any data to say that bushwalkers are irresponsible. How many "hoons" do you see bushwalking?
- Is the bush in the RPZ pristine – need to keep non-indigenous animals and plants out too. Feral pigs make more mess than bushwalkers.
- Surely any erosion from walking along a designated track close to water can be managed, though toileting is an issue.
- "Freedom to roam" has a good philosophical base and would be good in an ideal world. But when does it become a problem? Limit numbers and time in the RPZ and forbid toileting?
- 99% of recreation is fine and people mostly keep to tracks and designated camp sites, especially on the Bibbulmun and Munda Biddi Tracks. However, there is a lack of understanding between DEC and DoW as to what constitutes a camp site so there are not that many designated camp sites. But they are well maintained. Most are outside the RPZs except around Mundaring due to the RPZ only recently being declared there.
- Organised groups generally follow the rules and apply for permits to hold events and bring their own toilets if there are a lot of them. But this then causes "residual use" when hundreds of individuals then copy and do the same things but without permits. So the groups get disheartened.
- Recreators have internal conflicts – both within groups and between groups.
- In the past there was an MOU between the WC and the Federation of WA Bushwalkers that allowed them to walk in the RPZ subject to conditions which could be checked by a Ranger. But the bushwalkers wanted more freedom than provided in the MOU and not just walk but camp in the RPZ. Also, Rangers' instructions and authority were not accepted by the bushwalkers.
- Most of the Munda Biddi Track people have a good understanding of the environment, but the track goes close to a waterfall and we've found people showering in the waterfall.
- Recreators want open fires, but this is dangerous given wildfires.
- Being proactive in relation to trail bikes: Joint Agency Implementation Committee (DSR; DEC; Local Govt; WC) looking at trail bike use – not just finding areas but trying to manage and educate them in safety and limiting catchment damage.
- Trail bikers are cheeky and hard to catch.
- We spend a lot of time working with recreation groups and explaining the issues, but it's the little groups and the individuals who are hard to get to.

ECOLOGICAL ISSUES

- There is probably a parallel between the RPZs and the "not take zones" in marine conservation. Recreational and commercial fishers challenge the validity of "no take zones". But scientists say they are necessary for breeding.
- If recreational fishing were allowed in the DW dams, there would be pressure to populate with introduced species as the native fish are not attractive to fisher people.
- If motorised craft were allowed on the DW dams, there would be more likelihood of foreshore erosion.
- It seems reasonable to assume that if you have undisturbed forest, you are likely to find remnants of endangered species there – especially in the riparian zone where there is access to water.
- From an ecological point of view, there is always value in having undisturbed areas. It becomes a trade-off between ecological values and social values.

- Even if there is disturbance from feral animals, there is still reduced disturbance overall if people are excluded. But how does the disturbance from feral animals impact on the life habits and feeding patterns of endangered species? They could live together if they are not competing for food and resources.
- Nowhere is pristine, but the RPZ is relatively undisturbed so don't disturb it further.
- Mammals would flourish if the zones were fenced off, but it is not feasible to fence all the RPZs. Too resource intensive to fence and maintain.
- Feeding of migratory birds is disturbed by the presence of people. They only have a small window of time to feed and put on weight before their flight back to the northern hemisphere. It only takes one human an hour to walk by to disrupt the feeding.
- Wetlands on the Swan Coastal Plain are becoming drained and so the migratory birds will be looking for other places in the future. But they need special circumstances e.g. certain water depth so they can get to the bugs – can the dams provide this? Seasonal protection is needed.
- Mammals have different roaming ranges but the need for their protection is probably year long.
- Endangered species are being pushed into smaller areas with the drying climate (stressed trees and food and lack of water). They need all the breaks they can get.
- Research has shown that there are ecosystem services and ecological values that also have social value.
- There is a feral pig control program – trap and shoot and we wouldn't want people walking by while a ranger is shooting a trapped pig. We shoot about 200 a year in the summer months which is helping to reduce the numbers. But shooting pigs is another recreational pursuit (illegal) and the shooters bring pigs in from the SW. We are working with Murdoch University on ways to control the pigs.

FUTURE ISSUES/CLIMATE CHANGE

- With climate change and reducing rainfall, the nature of the dams will change, but exactly how, we don't yet know. For example, lower water levels, different in-fill rates, more fire risk and subsequent contamination and sedimentation, population pressures, different sun and heat leading to algal blooms etc.
- There are water reserves in the SW and there needs to be an investigation of which might be required for drinking water in the future, given the drying climate.
- There will be a move to alternative technologies (recycling, desalination) and hence a reduction of importance on the dams' catchments.
- Some dams will retain their importance. The desalination plant has to run 24/7 and it pumps to Wongong when not needed for immediate supply, so this dam will have to be kept clean if clean water is going in. Mundaring supplies water to agriculture and the Goldfields – already adding a filtration plant for Mundaring because of Kalamunda and Helena.
- All sources aren't equal. With climate change Water Corp is forced to go to other technologies and sources (e.g. recycling). So we need to have at least one pristine source to fall back on if other sources have problems.
- The future will bring greater international requirements for DWQ standards and the usefulness of the dams will go down. The cost of alternative technologies and treatment of water from the dams will become more comparable.

EQUITY AND COST

- It's a man made system, how should the cost of additional treatment be distributed. For example, why shouldn't the recreators have to contribute to the maintenance of the dam wall at Waroona.

GENERAL

- Why go backwards from a proven system?
- We don't have enough data to answer some of the questions yet.
- There is a lack of scientific evidence of the effects of recreation, and this is hard to get as it happens over a long period of time. Recreators use this argument. But the reviews show the damage. Other land uses, such as mining and harvesting are heavily regulated and you can't regulate recreation to the same extent.
- In America they are gradually taking DW dams back from recreation because of the increasing cost of treatment. There have been some pathogen related deaths.
- The community needs source protection education.
- But if you raise the profile of the hills catchments, there is a danger that more people will want to go there.

APPENDIX 2
OUTCOMES OF SCOPING FOCUS GROUP WITH GENERAL COMMUNITY IN WA

BACKGROUND

Sixteen stratified random community members were recruited to attend the discussion group. Four people were recruited from each of four regions in the Perth Metropolitan area: coastal; central/inland; eastern; and hills. Fifty percent males and females were recruited with attempts to provide a mix of ages under and over 40, and higher and lower socio-economic status. Professionals in the area of health industry, environmental science or conservation were not eligible.

Twelve community members participated on the night, six males and six females, and they were assigned to one of two groups (predominantly hills residents and predominantly coastal residents). Two members of the Water Corporation attended to provide information and answer any questions. Each group was assigned a facilitator to lead a series of discussion questions, and proceedings were digitally recorded.

Brief background information was provided in a presentation by the Water Corporation which covered the following key facts:

- What is a catchment and what is the current policy of catchment protection?
- How is water currently treated and delivered from the dams to households?
- How are catchments managed and water treated in other places nationally?
- What are the health and environmental implications of opening catchments to recreation?
- What are the treatment implications of opening catchments to recreation?

The following briefly outlines the key points from the discussion under each question heading. However, the discussion was free-ranging and there was considerable overlap of issues between questions.

Q1. Have you been aware of this policy previously? – What do you think are the good aspects .... and the bad aspects?

- Most indicated that they had been generally aware of the policy.
- Good aspects.
    - It's not a policy, it's common sense.
    - It avoids high cost of treatment.
    - We pride ourselves on safe drinking water, so we shouldn't jeopardise this.
    - Now I know what goes into the water from opening catchments, I don't want them opened.
    - The most important aspect is the safety of the water.
    - People can drive 100kms to other places.
    - We've ruined our food with chemicals, we need to keep the purity of our water. We don't want the added chemicals with the added treatment.
- Bad aspects
    - There was nothing forthcoming in terms of bad aspects of the policy.

Q2. If you were in charge of the policy would you leave it as it is, or would you change it? Why? ...... How?

- There should be no compromise for the policy. The only possibility is that maybe the 2km RPZ could be changed to 1 km.
- Wouldn't change it. It's a public good. Minority groups have too much influence.
- Difficult to get permits for some activities outside the catchments now.

- The longer we can hold off letting people in, the better it will be for our children.
- How many who are fined for entry to the RPZ are repeat offenders?
- Use the money saved on extra treatment to employ extra Rangers.

Q3. What do you do for recreation? What sorts of things do you do? Where do you do it? Would you benefit from the catchments being opened up to more human activity?

Following is not an exhaustive list, but provides the range of recreational activities pursued by the participants.

- **Water ski** – it's becoming harder and harder to find places to water ski. Too crowded on the river and not allowed to ski in the sea. Perhaps special areas for skiing could be built and the users pay? Would be happy to go to irrigation dams rather than drinking water dams.
- **Race motorcycles** – areas are being closed down and the strict management of other areas makes it difficult to organise events.
- **Bushwalk the Bibbulmun Track** – use the designated camp sites – happy to stay on the track.
- **Walking and camping down south** – there are some great places there.
- Prefer the beach.
- **4WD and camping** – mostly in National Parks. Never had a problem with where to go.
- **Fishing (in a club)** – SW rivers and beaches. If there are restrictions they are there for a reason and must be enforced. Fisheries inspect catches all the time.
- **Camping, fishing, travelling, beach** – before I came tonight I wanted catchments opened but with some education tonight I've changed my mind. I thought it was public land and I didn't know the water quality was as good as it is. I thought it already had extensive treatment with lots of chemicals, but the simple treatment now is best. There are other places to go. We are not an over-populated state – there's plenty of land and few people.
- **Visit local parks** - there are plenty of local parks when you get out and about in the city.
- **Family outings to picnic and play around water bodies** - there are plenty without the dams.

Q4. Some people totally agree with the policy and believe catchments should be kept pristine for total safety of drinking water.

Some people believe that a few more activities should be allowed closer to the water (e.g. bushwalking or horse riding), but not in the water ----- or perhaps allow no more contact with the water than fishing from the edges.

Some people want the catchments totally opened up for all activities and have greater treatment to ensure safe drinking water.

Thinking of these three points of view, what are the advantages and disadvantages of each. Which point of view do you mostly agree with?

- Agree with the first point of view and we know the disadvantages.
- This is an issue for better planning for recreation. Drinking water shouldn't be compromised.
- People are unaware of the many places available such as Lake Leschenaultia, which is really big and supports a lot of recreation.
- It's not good to swim in fresh water.
- An advantage of opening the catchments is that it would be good for tourism.
- Dam walls are interesting for tourists and they have access to those.
- If you could guarantee responsible behaviour, you could open up the catchments but you won't always have responsible behaviour. People will always break the rules.

- Keep it all the same, including the policing. If you draw too much attention to protection, it will create more demand and more people will be on the catchment and there will never be enough Rangers.
- It's not just a matter of opening up the catchments as there is conflict between passive and active recreation. If you allow one, it's only fair to allow the other, and this puts drinking water at greater risk.
- Bushwalkers don't like keeping to tracks and they tell others of their off-track experiences and the demand and subsequent impact gets bigger and bigger.
- We didn't know about fire risks, and that would be increased through recreation. There are probably a lot if issues the public aren't aware of that we have to be cautious of, such as fertilisers and other pollution.
- Need the information we were presented tonight into the schools. Put it on TV in short commercials.
- In 50 – 100 years time, what will our children be drinking?

Q5. There are some who say that those who want to recreate in the Reservoir Protection Zone (RPZ) can find other places instead.

Others say that with increasing population it's harder to find places to go that are close to where they live, and recreation is important for good health.

Are there other places that people can go? What are the health trade-offs – healthy lifestyle/healthy drinking water?

- There's the ocean.
- People aren't aware of the extent of Perth's parkland, just for children alone. I didn't know until I had children.
- People aren't aware of the extent of bike trails that are available throughout Perth.
- Water Corp should send out this information in their Newsletter.
- People don't want to go to the catchments because there is nowhere else to go. They are going there for the remoteness and wilderness experiences that aren't available at other places. But if the catchments were opened up, those experiences would be gone because of all the people that would go there.
- Perhaps there could be a ballot system like there is at Rottnest Island.
- Areas for off-road vehicles are being closed down. There needs to be better planning for recreation.
- Alternative places need to be found because once people go the catchments a few times (even illegally), they think they have a right to do so in the future.
- Is the land as fragile as we think?
- Is there a difference in environmental condition between the 2km zone and the 5km zone?
- Perhaps 2kms is too close.
- We are giving the city perspective. What about country towns and rural people? It might be there only form of water based recreation.
- The dams are not just attractive to international or interstate tourists, they have tourism potential for rural people.
- There are sacred sites such as in Arnhem Land with limited access. We accept those, why not accept catchment restrictions.

Q6. It is said by some that there are costs to not using the land
- development and tourism loss
- fewer recreational opportunities and so increased health costs if it is harder to find other places to recreate

It is also said by some that there are costs to using the land
- loss of pristine land: threats to biodiversity; loss of protected habitat for endangered species; environmental damage such as erosion
- increased fire risk and hence potential to contaminate the water
- increased financial costs: treatment infrastructure; more sophisticated treatment; greater energy costs associated with the greater treatment.

What is important to you if you had to make the decision about using or not using the land?

If the land was opened up, who should pay for the management and treatment?

- Most important is the quality of the water combined with the cost – not just the financial costs, but the health costs of additional treatment.
- Purity of water is most important. It's hard work keeping "crap" out of our kids mouths now, but at least we can give them healthy water to drink.
- There is a natural filtration process and it retains the minerals we need for health.
- The dams are not big and so the area we are locking up is minimal.
- Tourists come to WA for our natural attractions and it's a big state, and we are only keeping them out of the dams and a small amount of land. The hills aren't an international tourist attraction anyway.
- Who pays?
    - Public pays – we all pay eventually.
    - Should be user pays.
    - Even if we started with user pays, it would end up with everyone paying.
    - It's not about cost, it's about prevention.
    - What are we willing to pay: financial, environmental, health. We always wait until things are critical before doing something about it. Prevention is better.

APPENDIX 3
OUTCOMES OF SCOPING FOCUS GROUP WITH RECREATORS FROM FORMAL GROUPS IN WA

BACKGROUND

Sixteen recreation groups representing a mix of land and water, passive and active, recreational pursuits were invited to participate in the discussion.

Eight groups accepted the invitation while another two apologised that they couldn't attend. The eight groups represented:

- Bushwalking (2 representatives)
- Rogaining
- Orienteering
- Freshwater Fishing
- Trail Bike Riding
- Water Skiing
- Boating

Two members of the Water Corporation attended to provide information and answer any questions. The discussion was facilitated in one group and attempted to follow the same series of discussion questions as occurred in the Community Group. Proceedings were digitally recorded.

Brief background information was provided in a presentation by the Water Corporation which covered the following key facts:

- What is a catchment and what is the current policy of catchment protection?
- How is water currently treated and delivered from the dams to households?
- How are catchments managed and water treated in other places nationally?
- What are the health and environmental implications of opening catchments to recreation?
- What are the treatment implications of opening catchments to recreation?

The following briefly outlines the key points from the discussion. As the discussion was "spirited", little of the originally designed structure occurred. However, all points were covered at some stage.

PRIOR RIGHTS

- The Federation of the WA Bushwalkers had written permission from the Water Corporation to walk in the 2km RPZ until about 10 years ago when it was taken away. They were then the eyes and ears of the Government. Rangers were advised in advance where, when, and how long they would be there. There was no impact on water quality.
- There was an agreement up until the 1980s for trout fishing in the top of Serpentine dam.
- In the 1980s orienteering had the use of catchment areas as far as the high water line. That became the 1km RPZ which about 10 years later became the 2km RPZ.
- Policy 13 said that historical agreements would be honoured but they weren't.

INDIVIDUAL GROUPS' PERSPECTIVES

BUSHWALKERS: state that catchments are full of pigs that pollute similarly to people. Mining is allowed, along with forestry and monocultures. You can mine it and log it but you can't walk in it. As long as you don't access the water and follow good practices, there is not a problem. There is no scientific evidence to say that the 2km exclusion zone works. Should start with "freedom to roam" and work from there with appropriate management and infrastructure.

TRAILBIKE RIDERS: are happy to concede the RPZ with the exception of sealed roads. But they don't like not being able to use the outer catchments with the assumption that they can't be managed. They believe the risks to water quality are exaggerated and that exclusion fails to recognise management. There should be management and not exclusion. The outer catchment boundaries are lines drawn on maps and not where the water flows.

FRESHWATER FISHING: believe there are far more "natural" pollutants in the catchments than they would produce. The allowable areas for fishing are confusing, especially in the streams. The signs stating what is allowed and where are confusing and don't indicate where the water is running to.

ROGAINING: have organised events, about 6 year with about 400 people at each event. They feel they have little impact on water quality. They have to interact with the Water Corp (WC) and DEC and try to follow the rules. But there are conflicting advice and statements from the two agencies – often at cross-purposes. It's not a friendly interface. They accept that water quality needs to be protected, but there must be balance. It can't be compared nationally as there are different climates and terrains.

ORIENTEERING: have similar issues to rogainers. They have 20-25 events a year, of which 15-20 are held in catchment areas. Their season is winter. They believe that all the disadvantages of the regime are aimed at organised groups. They negotiate with WC and DEC for an event in an area and when they turn up they find individuals there breaking the rules and not being policed. Only the organised groups are policed. The best areas for orienteering require detailed topographic terrain and in rock and granite areas. Most of these areas are in the northern catchments. The WC and DEC ask why they don't go down to Dwellingup, but it's steep, thick and dense bush, and no terrain. They rely on the Helena and Canning catchments. A government agency, Healthway, provides about half their funds to promote good health, while other Government Departments say they can't do this or that.

WATER SKIING: members probably have the greatest impact on water. The association has been going for 50 years with about 500 members. Including the social skiers, there are probably 5,000-10,000 people in WA in the sport. Mostly ski on the Swan and Canning Rivers and use rural lakes and dams and inland waterways. They need as much water as they can get. They work well with DoT and CALM. When Logue Brook was (temporarily) closed, it had a big impact on water skiing. Taking aside natural pollutions and dead animals, many national and international studies have shown, that with modern boats, there is very little pollution in the water from boats.

BOATING: the organisation has been around for 11 years with 1,600 members. They mostly boat on the Swan and Canning Rivers and up and down the coast. However, it's a very nasty coast for sailing and it would be ideal to have access to inland waters. Modern technologies mean far less pollution from out-board motors.

INDIVIDUALS, GROUPS AND POLICING

- There used to be designated areas for Trailbike riders. There are now only 4 out of 7 areas left. Less ability to run organised events now with the reduced areas. So by penalising the groups, the activity disperses and you lose control over it.
- Trailbike Assoc doesn't run events, but because the bikers aren't allowed to be on the catchments, just by being there they are breaking the rules. So one bad behaviour is no worse than another. What is missing is a sense of privilege in being on the catchment and a sense of custodianship for the catchment. If they are told they can't be there, there is no feeling that this is a place to be looked after. Individuals need to be convinced that it is their responsibility to care for the catchment.
- Trailbike riders need trails as well as areas. Need more areas for those who just want to ride around in circles and allow the destination riders to use the trails.
- Jet skis have fewer and fewer places to go. Far more people in the sport than there are places for.
- The more people that are in the area, the less likely people are to dump cars, light fires and pollute etc. They tend to look for more isolated areas.
- If you have to walk into an area, and not just drive up to it, there is less likelihood of "yahoos" in the area.
- It needs to be self-policing. Boaters can't discharge waste to the ocean and they are required to have holding tanks and that works quite well. So self-policing works.

- Self-policing doesn't work. The number of unorganised recreational activities on the catchment bush is increasing.
- There'll always be people who want to buck the system
- Rangers are paid from Monday to Friday but it's the weekends when policing is needed for individuals.

MANAGING DEMAND, ALLOCATION AND ARE THERE OTHER PLACES?

- There is a limited amount of prime quality land and there are a large number of recreational users and uses and this will increase in the future with increasing populations and so the space actually becomes smaller.
    - It's a spatial and temporal problem.
    - What activities can co-exist and when, and when should others go elsewhere?
    - Need a master plan, a multi-objective approach, to maximise the utility of the area rather than only maximising the water quality.
        - How best to utilise this land for recreation as well as protecting the water.
- Economics over-ride the prime objective of protection of water quality i.e. mining overrides water quality as well as safety with controlled burning. But recreation also has an economic benefit.
- Groups have to work together to hold water events on the Swan and Canning and about 500 events are gazetted on the calendar each year. The same could be done in the catchments and all groups would need to work together with the government departments to make it work.
- Need better access to get people in and out, especially if there is an injury. Bushwalkers are allowed to go in more places than others, but it's hard to get there because of closed off car tracks etc. Family bushwalkers can only walk within 3 kms of a car park.
- Need a shared resource trail head - one shared parking facility with different trails going out from it (e.g. mountain bikes, trail bikes, walking, 4WD etc.). With more people around it would help stop break-ins,
- Need better and more toilet facilities and camps in the places where they are needed.
    - Groups have to take their own chemical toilets to events.
    - Bushwalkers rely on the use of best techniques to dispose of faeces on their walks. They dig holes and rely on the assimilation of the environment.
- Orienteering would need the resources close to their map.
- Rogainers have had to find other areas in the past and will find them in the future.
    - Up to 60% of activities are on private property or farms.
- Artificial water bodies could be built.
    - Champion Lakes has virtually taken 90% of the rowing off the Swan and Canning Rivers, so other lakes could do the same for water skiing and boating.
- Kings Park is good for bushwalking, but the controlling agency is very restrictive.
    - Walking areas in Perth are full now – so many people are out in the evening.
- Would more areas actually draw from demand or add to demand?
- Recreational areas are being reduced because of threats of litigation from accidents

MANAGEMENT ISSUES

- Need one coordinated body to deal with rather than several government departments with conflicting advice and requirements.
    - Used to have that with the Water and Rivers Commission who formed a multi-agency committee to make the decisions on activities.
- Look for international models for management and education.
- There's a lot of misinformation that is affecting people's concern for DWQ. Recreation is not that big a risk on water quality if it is managed responsibly.

COSTS AND BENEFITS

- The aesthetic value is paramount – to go to the top of a peak and look out and just see forest.
- Nature does ultimately prevail. What we do to the land is only temporary. To preserve forests, stop clearing and wholesale damage. But do we need to treat forests so gently in order to preserve them. Will a few people destroy them?
- Power lines are coming and bauxite mining is coming in the future. There's noise pollution from mining. But mining does rehabilitate – though not always back to original forest type and condition.
- Landscape is important. Keep to recreation rather than other developments.
- General public feel shock and horror about boats and skiing on the water, but it's the pollution from the land that is the issue.
- Once mining is finished, so is the economic good. But the economic benefit of recreation is ongoing.
- The carbon footprint is reduced when driving to the hills rather than to the SW.
- A lot of SW towns rely on recreation economically.
    - Food, fuel, fishing, facilities etc.

WHO SHOULD PAY?

- Apart from direct recreational and economic benefits, there are social benefits from recreation that flow on to the community e.g. a socially stable community. Therefore society can pay. That's why we have taxes and social welfare.
- Before the dam, the bush was there for use, but now the suggestion of pay for use. Remove the dam and there's nothing to pay.
- Pay indirectly to allow recreation or pay for more health treatments.

APPENDIX 4
REPORT OF THE SURVEY PHASE OF THE WA CASE STUDY

The survey area and drinking water dams and catchments are shown in Figure 4.1.

METHODOLOGY

The questionnaire was administered by two different processes:

- personal telephone interviews of the random community sample and the recreator sample; and
- web survey of the professional sample.

Both processes were conducted by a local data collection company to ensure confidentiality of responses from the researchers who recruited the recreator and professional samples. The telephone interview averaged 28 minutes in length and required considerable input by respondents.

COMMUNITY SAMPLE

A target stratified sample of 600 random residents was set for the general community survey. This was to be achieved through 75 responses from each of eight geographic locations. Locations were defined by relative distance from the drinking water catchments, and separated by coastal and inland. The assumption was that people closer to the coast had more easily accessible alternatives to the drinking water catchments for their recreation pursuits. Half the sample were to be male and half female, and half under 40 years of age and half over 40 years of age. All respondents were to be over the age of 18 years.

The target numbers and achieved numbers for each location are shown in Table A4.1 below and a total sample number of 604 was achieved. The gender and age targets were achieved exactly.

Table A4.1. Details of the Community Sample

Location	Target N	Achieved N
Perth Metropolitan Area		
Far from catchments - inland	75	75
Far from catchments - coastal	75	75
Closer to catchments - inland	75	77
Closer to catchments - coastal	75	75
Near to catchments - inland	75	77
Near to catchments - coastal	75	75
South-West		
Near to catchments - inland	75	75
Near to catchments - coastal	75	75
Total	**600**	**604**

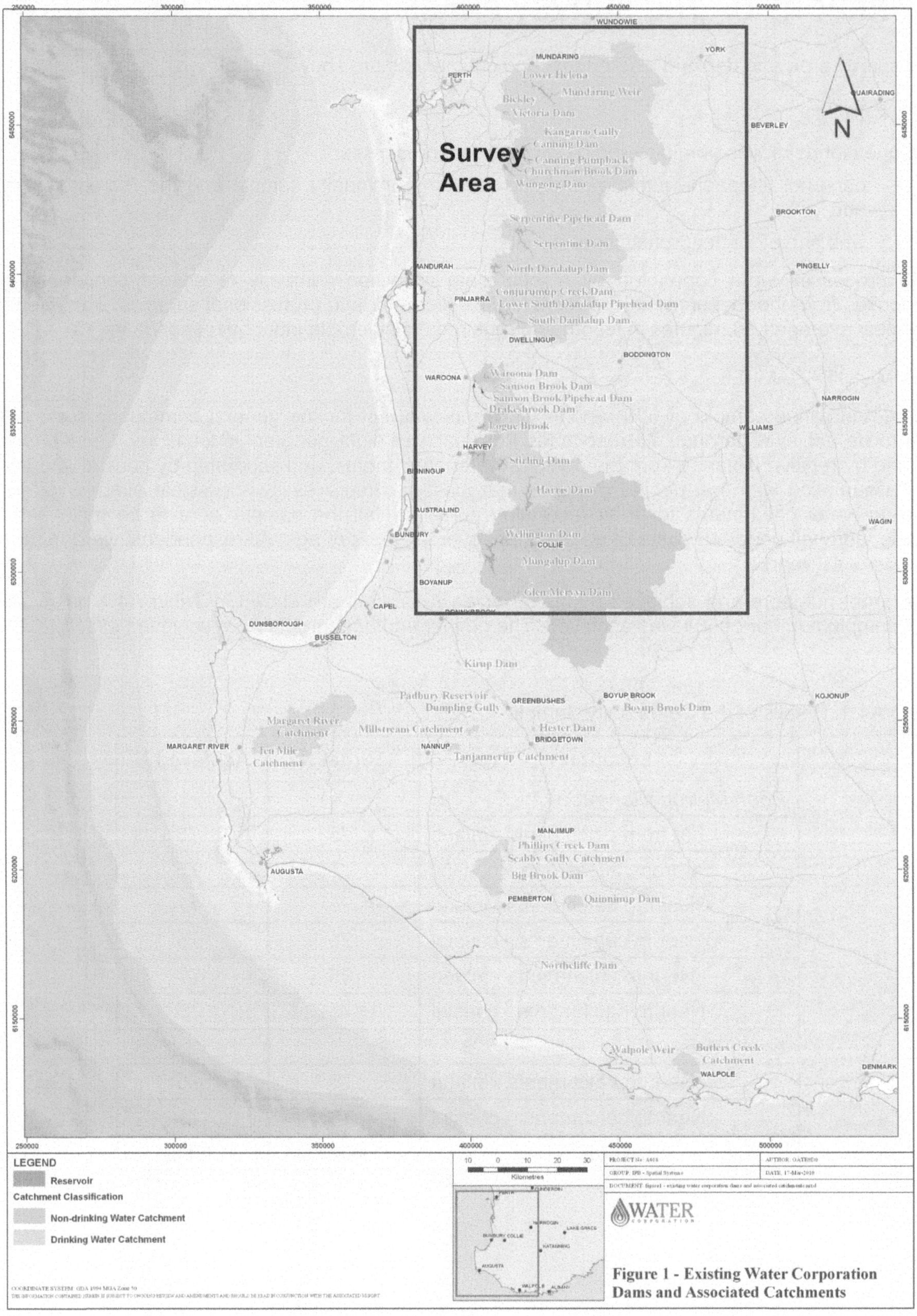

Figure A4.1. Survey Area and Drinking Water Dams and Catchments

The refusal rate for the survey was 78% which is consistent with surveys of this length. While some may find this refusal rate of concern, there is a distinction between response rate and response bias. Research has shown a generally poor correlation between survey non-response and response bias (Groves, 2006). That is, just because there is a low response rate, this does not necessarily equate to biased data.

RECREATOR SAMPLE

While it was hoped to attain a sample of 100 recreators who were members of formal recreation groups, the emphasis was on obtaining a range of active and passive, and land-based and water-based activities. The goodwill and cooperation of the groups was required to obtain lists of people and their contact details who agreed to be called and interviewed. A four wheel drive association and motor boating organisation were also invited to provide people to participate in the survey, but did not respond in the time. However, the water skiing people involve motor boating, and there was a good representation of trail bikes and motor bikes. The bushwalkers represented three different organisations. A total of 68 recreators agreed to participate. Table A4.2 shows the numbers from each activity that were provided for contact and the actual interview numbers achieved. The final response number of 66 was sufficient to allow valid statistical comparisons. The survey data base did not identify respondents by recreation group to ensure confidentiality as numbers in the groups were quite low.

Table A4.2. Details of the Recreator Sample

Group	Agreed N	Achieved N
Bushwalking	7	7
Orienteering	10	10
Rogaining	11	11
Trail biking/motor cycling	13	13
Freshwater angling	10	10
Canoeing	10	9
Water skiing	7	6
Total	**68**	**66**

PROFESSIONAL SAMPLE

A range of professionals of different relevant disciplines and from the public and private sectors were invited to participate in the survey through a web based format. This format was chosen for the professionals due to the length of the survey and the fact that they were completing it in the course of their work. This was unlike the other two samples who were telephoned at their homes at times of their convenience. Using this web format, it allowed the respondents to answer the questions in one sitting or in several sittings, to fit with their work schedules.

A minimum target of 30 was proposed and 43 were invited to participate. A networking process was used to identify people working in the area in the required range of disciplines. They were invited to participate by email and provided with a copy of the project proposal. Thirty-six agreed to participate when invited and in the end, 28 actually completed the survey.

Table A4.3 shows the number from each organisation who agreed to participate and the number who actually completed the survey.

The two Department of Sport and Recreation professionals, one consultant and one Department of Health professional advised that they would not complete the survey due to their assessment that the questionnaire was biased.

The survey data base did not identify respondents by organisation, to maintain confidentiality.

Table A4.3. Details of the Professional Sample

Organisation	Agreed N	Achieved N
Consultants	3	1
CSIRO	3	3
Curtin University of Technology	3	3
Murdoch University	1	1
Forest Products Commission	3	3
Tourism WA	1	1
Department of Health	3	0
Department of Environment and Conservation	4	3
Department of Sport and Recreation	2	0
Department of Water	6	6
Water Corporation	7	7
Total	**36**	**28**

Table A4.4 shows the disciplines represented in the sample as recorded by the respondents, some being more specific than others.

Table A4.4. Disciplines Represented in the Professional Sample

Discipline	N
Research Scientist/scientist/science	6
Water sciences/management/engineer	5
Planning/social research	4
Environmental science/management/policy	3
Forestry science/management/operations	3
Recreation/tourism	3
Chemistry	2
Microbiology	1
Management	1

COMMENT ON THE SURVEY SAMPLES

While the professionals survey sample may be considered too small and disparate for analytical results to be meaningful, we make the following observations.

Firstly, as a sample of professionals with a varied interest in recreation, drinking water catchment management and drinking water treatment and supply, the rationale for this sub-set is no different from that of the recreator sample or the community sample. The pool of professionals in this area is not large and therefore a sample of 28 is defensible. The disparate nature of the professional disciplines represented is not unlike the disparate nature of the activities represented in the recreator sample, or the range of views offered by a random community sample.

While we acknowledge that the small number in the professional sample lacks statistical power in that statistically significant differences can be more difficult to show, key differences between the professionals and the community did occur, and major differences occurred between the professionals and the recreators.[15]

[15] A conservative approach to reporting statistically significant differences has been taken ($p < .01$ rather than $<.05$) to ensure that the differences that occurred were robust.

THE QUESTIONNAIRE DEVELOPMENT

The questionnaire was designed based on the first phase of the project: the qualitative data collection and a scientific literature review. The design of the questionnaire was also assisted by the national Project Steering Committee and the local project Working Group.

A near final version of the questionnaire was pre-tested on 24 respondents from the eight sampling regions and changes were made. Interviewers were specifically trained on the final questionnaire and provided with background information, and specific written information for respondents who might ask questions or required greater clarification of issues.

The recreator and professional samples were provided with the opportunity for additional comments as it was assumed that these respondents, having a close interest in the topic, may not be provided with sufficient opportunity in the set questionnaire to impart the information that they were particularly interested in. The individuals in these two samples were specifically recruited for the survey and so it was possible that they might have had a number of points they wished to make, and the opportunity to provide additional comments ensured that this could occur.

The additional comments are detailed in Addendum 1 and 2.

Results

This section outlines the results of each question in the survey with comparisons of responses between the three samples, the data reduction analyses, the multivariate analyses, and demographic comparisons. Oneway analyses of variance were conducted to compare responses to questions between samples and key demographic groups and where statistically significant differences occurred ($p<.01$), these are explained.

PARTICIPATION IN RECREATIONAL ACTIVITIES

Respondents were firstly asked if they participated in any form of outdoor recreation. The majority of the community sample, 84.3% (N=509), replied in the affirmative with only 15.7% saying they didn't. All of the respondents in the recreator and professional samples said they participated in recreational activities.

Those respondents who answered yes, were asked to give details of their recreational activities. Responses were unprompted for the community and recreator samples, but activities were provided to professionals in the web survey. Table A4.5 provides the details for each sample under general headings. As respondents were able to provide more than one response, the percentages are of the number of cases in each sample number and do not add to 100. Most of the community and recreator samples participated in up to four activities, with 21% of the community sample and 10% of the recreator sample participating in more than four activities. Not surprisingly, as the professionals were responding to options that were provided due to the different method of questionnaire delivery, 68% said they participated in more than four activities.

Table A4.5. Details of Recreational Activities for the Three Samples

Activities	Community % N=509	Recreators % N=66	Professionals % N=28
Land-Based Passive			
Walking	67.2	21.2	71.4
Cycling	20.4	24.2	46.4
Picnicking	18.1	13.6	60.7
Camping	10.2	19.7	46.4
Running	7.1	12.1	14.3
Bushwalking	4.3	21.2	60.7
Hiking	2.2	6.1	25.0
Mountain Biking	1.2	6.1	25.0
Climbing	0.6	3.0	7.1
Orienteering	0.4	16.7	0.0
Rogaining	0.0	16.7	0.0
Other	5.9	0	3.6
Land-Based Active			
4WDriving	3.5	1.5	21.4
Trail Biking (motorised)	1.0	24.2	3.6
Hunting/Shooting	0.2	0.0	0.0
Other	0.6	0.0	0.0
Sport Requiring Specific Facilities			
Swimming in a Pool	7.1	1.5	42.9
Football/Soccer/Rugby	2.4	0.0	14.3
Tennis	2.4	1.5	17.9
Cricket	1.2	0.0	10.7
Basketball	1.0	0.0	3.6
Hockey	0.8	0.0	3.6
Netball	0.6	0.0	3.6
Other	5.9	0.0	17.9
Water-Based Passive			
Fishing	25.1	24.2	42.9
Swimming	16.5	7.6	60.7
Canoeing/Kayaking/Rowing	3.5	28.8	35.7
Sailing (no motor)	3.3	1.5	7.1
Board/Body Surfing	2.4	1.5	39.3
Wind/Kite Surfing	1.0	0.0	3.6
Paddle Boarding	0.4	1.5	0.0
Other	4.9	1.5	3.6
Water-Based Active			
Motor Boating	6.9	4.5	7.1
Water Skiing	3.1	10.6	14.3
Jet Skiing	0.2	0.0	3.6
Other	1.0	0.0	3.6

Interestingly, recreators were less likely to nominate sporting activities that required special facilities than the other two samples. Respondents who nominated any activities other than "Sport Requiring Specific Facilities" were asked where they mainly participated in the activities. Table A4.6 shows the general categories of locations where these occurred.

It was evident that recreators participated considerably more in *rural or bushland areas* than either of the other two samples and were also more likely to participate at *lakes and dams* than the other two. The community and professional samples more often participated in *urban, city or town areas* and *oceans and estuaries* than at other locations.

Table A4.6. Where Recreational Activities Occurred

Activities	Community % N=444	Recreators % N=66	Professionals % N=28
Urban, city or town area	59.2	43.1	85.7
Ocean or Estuaries	48.2	21.5	60.7
Rural or bushland area	33.1	84.6	53.6
Rivers	24.8	40.0	50.0
Lakes or Dams	14.0	40.0	21.4

Respondents who nominated any activities other than “Sport Requiring Specific Facilities” were also asked to rate their agreement or disagreement with six statements (which were rotated to avoid order bias) about enjoyment of activities and where they recreated. They were asked to do this and on a five point Likert scale from 1 being strongly disagree, through 3 being neither to 5 being strongly agree. The mean ratings for each sample are shown in Table A4.7 below.

Table A4.7. Mean Ratings of Agreement with Statements of Enjoyment of Activities and Places of Recreation

Statement	Community Mean Rating* N=444	Recreators Mean Rating* N=65	Professionals Mean Rating* N=28
I tend to go to the same few places for recreation **	4.02	3.51	3.25
I am involved in a wide range of recreational activities **	3.07	3.80	3.43
There are only a few places I'm allowed to go to enjoy my recreational activities **	2.45	3.66	1.93
More and more people are coming to my favourite places to recreate	3.54	3.71	3.25
My recreational activity is being banned in more and more places **	2.39	4.17	1.96
My recreational enjoyment is often lessened by other activities in the same place **	2.52	3.69	2.82

* The higher the number the greater the agreement with the statement.

** Statistically significant differences between samples (p<.01)

Of the community sample, 84% agreed (agreed or strongly agreed) with the statement that they tended to go to the same few places for recreation. There was a spread of opinion about involvement in a wide range of activities. Fifty-nine percent (59%) agreed that more and more people were coming to their favourite places while 20% were ambivalent about this statement. Of the remaining three statements concerning few places to enjoy activities, being banned in more and more places, and enjoyment lessened by other activities in the same place, while about two-thirds of the community sample disagreed with these statements, about one-quarter agreed with them. This indicates some discontent with opportunities to enjoy recreational activities of choice by the community sample.

Oneway analyses of variance were conducted to identify statistically significant differences (p<.01) between the three samples on the ratings of the statements. Differences that emerged follow.

- The community sample was significantly more likely to agree with the statement, *I tend to go to the same few places for recreation* than was either the recreator sample or the professional sample.
- The recreator sample was significantly more likely to agree with the statement, *I am involved in a wide range of recreational activities* than was the community sample.

- The recreator sample was significantly more likely to agree with the statements, *there are only a few places I'm allowed to go to enjoy my recreational activities; my recreational activity is being banned in more and more places;* and *my recreational enjoyment is often lessened by other activities in the same place* than was either the professional sample or the community sample.

The third point of difference by the recreator sample is consistent with the quarter of the community sample noted above that is concerned about opportunities to enjoy their recreational activities.

DRINKING WATER AND ITS PROTECTION

Respondents were asked if they knew where their drinking water came from. One-third of the community sample, one recreator and two professionals said they did not know. Those who thought they knew were asked unprompted to name the sources. They were allowed multiple responses. Figure A4.2 shows the main drinking water sources offered by each sample.

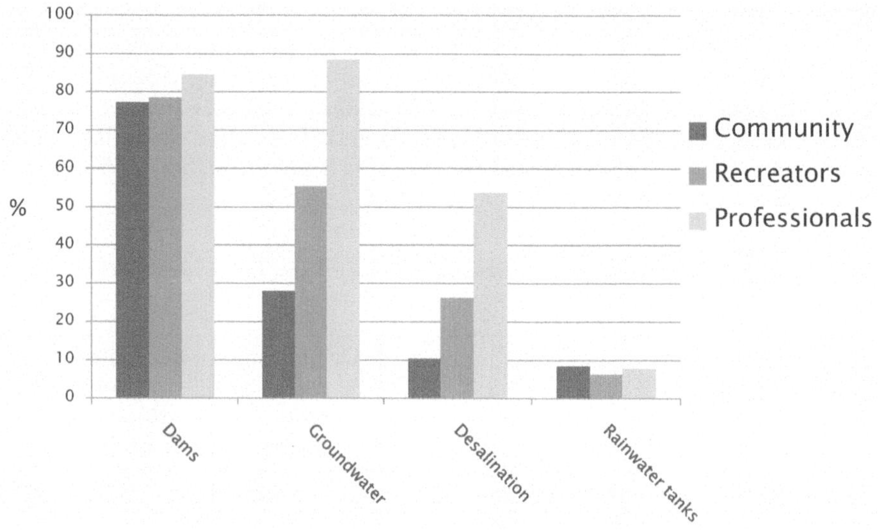

Figure A4.2. Knowledge of where Drinking Water Comes From

Fifteen of the community sample could not name a drinking water source and of the 3% who answered "other", seven said *bottled water.* Drinking water from the dams was most well known by the community and recreator samples with just over a quarter of the community sample and half the recreator sample naming groundwater. The lesser known source was desalination with one-tenth of the community sample, a quarter of the recreator sample and half of the professional sample nominating it.

Respondents were then asked if, to their knowledge, they had ever visited or recreated at a drinking water dam or catchment in WA. Seventy-three percent (73%) of the community sample said they had, with only 24% saying they hadn't. Nine community members didn't know. Eighty-six percent (86%) of recreators said they had visited or recreated at a drinking water dam or catchment in WA and 96% of professionals. No recreators or professionals were uncertain if they had.

Respondents who answered *yes* or *maybe* were then asked to name the dam or catchment they had visited. It is evident from the Table A4.8 below that respondents in all three samples did not necessarily discriminate between drinking water and irrigation dams.

Table A4.8. Drinking Water Dams and Catchments Visited for Recreation

Source	Community % N=449	Recreators % N=66	Professionals % N=28
Mundaring	55.9	62.1	82.1
Serpentine	25.6	24.2	42.9
Wellington	21.3	22.7	17.9
Canning	13.1	22.7	42.9
Harvey	11.6	12.1	7.1
Stirling	6.5	10.6	17.9
Harris	5.8	4.5	3.6
Waroona	5.1	6.1	3.6
Logue Brook	4.7	12.1	7.1
North/South Dandalup	4.0	7.6	39.3
Wungong	3.3	19.7	28.6
Kununurra/Lake Argyle	2.9	0.0	0.0
Churchman Brook	0.0	3.0	7.1
Bickley	0.0	7.6	0.0
Helena	0.0	6.1	0.0
Victoria	0.0	0.0	10.7
Harding	0.0	0.0	3.6
All Hills Dams	0.0	1.5	14.3

The most frequently visited catchment was Mundaring for all three samples. Serpentine was the next most popular dam/catchment visited. It should be remembered however, that a number of professional respondents visited these catchments in the course of their work and so will be over-represented at some locations. It is interesting that the visiting pattern for recreators was not dissimilar to those of the community sample.

The respondents who had visited the dams or catchments were then asked to rate, on a four point scale with 1 being *rarely*, 2 being *occasionally*, 3 being *often* and 4 being *a lot,* how often they had visited the catchments in recent years. The mean rating for the community sample was 1.69, with **48% saying that they *rarely* visited the catchments**. **Fourteen percent said they visited the catchments either *often* or *a lot***.

The community's frequency of visitation was significantly less often ($p<.01$) than for recreators or professionals, though again a number of professionals visit the catchments frequently in the course of their work. The mean ratings for recreators was 2.47 and for professionals 2.56. **Forty-six percent of recreators and 52% of professionals said they visited the catchments *often* or *a lot*. Only 14% of recreators and 11% of professionals said they *rarely* visited the catchments**.

Respondents were then asked to rate on a five point scale from 1 being *no knowledge at all*; 3 being *some knowledge* and 5 being a *high level of knowledge*, how much knowledge they had about managing dams and catchments to help protect drinking water from pollution and contamination. The mean ratings for each sample is shown in Figure A4.3 below.

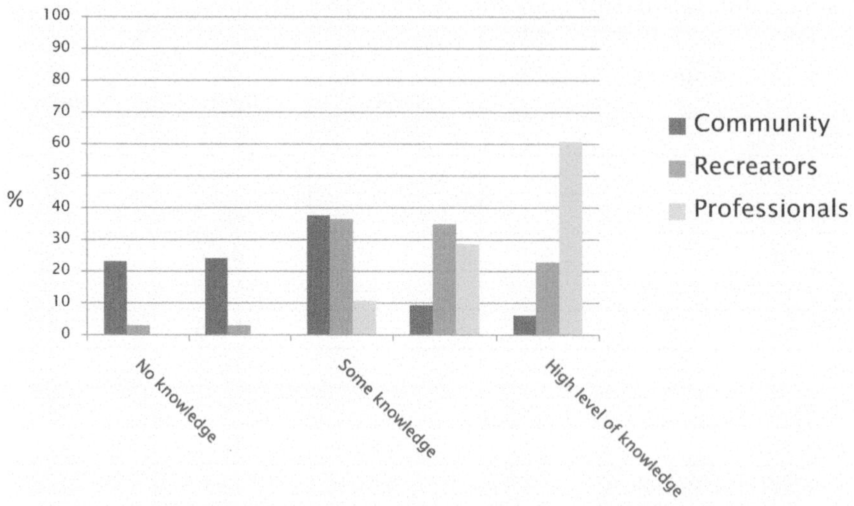

Figure A4.3: Mean Ratings of Perceived Knowledge about Drinking Water Protection

The mean community rating indicated a general poor perceived knowledge of drinking water protection with almost half the sample (47%) rating 1 or 2. Only 15% rated their knowledge as 4 or 5. Recreators perceived themselves to have significantly greater knowledge ($p<.01$) than the community with a mean of 3.71, and 58% rating their knowledge as 4 or 5. Professionals perceived themselves to be significantly more knowledgeable again ($p<.01$) with a mean rating of 4.50, and 89% rating their knowledge as 4 or 5.

Respondents who rated their knowledge greater than "no knowledge at all" were then asked, unprompted, what they thought was done to protect drinking water *before it was piped to treatment plants.* This question was asked to determine how much the respondents *actually knew as opposed to thought they knew.* These unprompted responses were then coded into categories as shown in Table A4.9. Many and varied answers were offered, some of which were correct and some were incorrect. Most could be coded into the broad categories shown in the table. However, some responses were offered by only a small number of people and so were coded in one of two "other" categories, depending on whether they were correct or incorrect answers.

It should be noted that the professionals tended to answer at a higher level and with less detail than the other two samples, with reference to legislation and regulation, but little specificity as to content. As these responses were many and varied, most were coded in the "other" categories and so represent a high proportion of professionals' responses.

Table A4.9. Unprompted Responses to What is Done to Protect Drinking Water

Source	Community %* N=464	Recreators %* N=64	Professionals %* N=28
Keep people away – not specific	21.7	60.9	28.6
Keep people and boats out of the water	19.0	43.8	14.3
Keep people from going near the water (e.g. 2 kms away)	15.3	51.6	32.1
Keep rubbish/chemicals/pollution away from the water	9.1	0.0	0.0
No motorised activity allowed other than driving on roads	8.6	20.3	3.6
Only allowed to picnic or camp in designated areas	7.1	28.1	0.0
Keep (feral) animals away from the water	6.0	0.0	10.7
Policing/signage	4.5	0.0	42.9
Only allowed to walk on designated tracks	3.9	25.0	3.6
Only allowed to drive on designated roads	3.7	18.8	3.6
Need government permits to hold large events	0.2	0.0	3.6
Other – correct	10.3	3.1	103.6
Other – incorrect	14.9	0.0	3.6
Don't know	29.3	6.3	0.0

* As more than one answer was allowed, percentages do not add to 100 and numbers of responses are percentages of cases.

A total *actual* knowledge score was calculated by assigning +1 to correct answers, -1 to incorrect answers and 0 to don't know and then summing scores for each respondent. Possible scores therefore could range from -6 to +6.

For the community sample, the mean score was .74 with almost half the sample (47%) scoring 0; 9% scoring -1; and 20% scoring 1. One person had the highest possible score of 6. This indicates a relatively poor knowledge by the general community about catchment protection for drinking water. A correlation analysis was performed between the *actual knowledge* score and people's rating of *perceived knowledge*. While the result was statistically significant, there was only a correlation of .36, indicating that there is only a moderate relationship between *perceived knowledge* and *actual knowledge*.

For the recreators' sample, the mean score was 2.44 with almost half the sample (47%) scoring 1, 2 or 3. No-one scored in the negatives and only 9% scored 0. Eight people (12%) had the highest possible score of 6. This indicates a significantly better knowledge ($p<.01$) than the general community about catchment protection for drinking water. A correlation analysis was performed between the *actual knowledge* score and recreators' rating of *perceived knowledge*. A similar correlation resulted as for the community, .37, and while it was statistically significant, this indicates that there is only a moderate relationship between *perceived knowledge* and *actual knowledge*.

For the professionals' sample, the mean score was 2.71 which is significantly greater ($p<.01$) than for the community but not significantly different from the recreators. Almost three-quarters of the sample (71%) scored 1, 2 or 3. One person scored -1 but no-one scored 0. Two people (7%) had the highest possible score of 6. However, as stated above, a number of the professionals responded with lesser detail than the other two samples. The correlation analysis between the *actual knowledge* score and professional's rating of *perceived knowledge* yielded a non-significant outcome of 24.

Respondents were provided with the following information about the current policy in WA to protect drinking water catchments.

> *There is considerable evidence that recreation increases the health and wellbeing of society. In WA there are many places available for recreation.*
>
> *However, the Government currently restricts recreation on our drinking water dams and catchments to provide high quality drinking water. Human contact with the water, and public access within 2kms of the water is not allowed. Activities such as walking and cycling are allowed in the catchments outside the 2km zone. Driving off-road, and any water-based activities such as swimming and boating, are not allowed anywhere in the catchments. This keeps contamination such as human bacteria and pathogens, chemicals, and sediment out of our drinking water, and maintains the natural processes that help to purify the water.*
>
> *Very little treatment is therefore required to keep our drinking water safe, and the cost to consumers is kept down. If the policy were changed and more recreation was allowed closer to or in the water, the risk of contamination would greatly increase. This would mean increased costs for <u>high level water treatment</u> and recreational management of about $600million. It is estimated that this would increase your water bill by about 15%.*

Respondents were then asked to rate on a five point scale how much they agreed with this policy. The mean ratings for each sample are shown in figure A4.4 below. There was considerable agreement with the policy by the community sample with 86% in agreement. Fifty-seven percent of professionals agreed with the policy while only 24% of recreators agreed with it. Recreators were significantly less in agreement than the professionals ($p<.01$) who were significantly less in agreement than the community ($p<.01$).

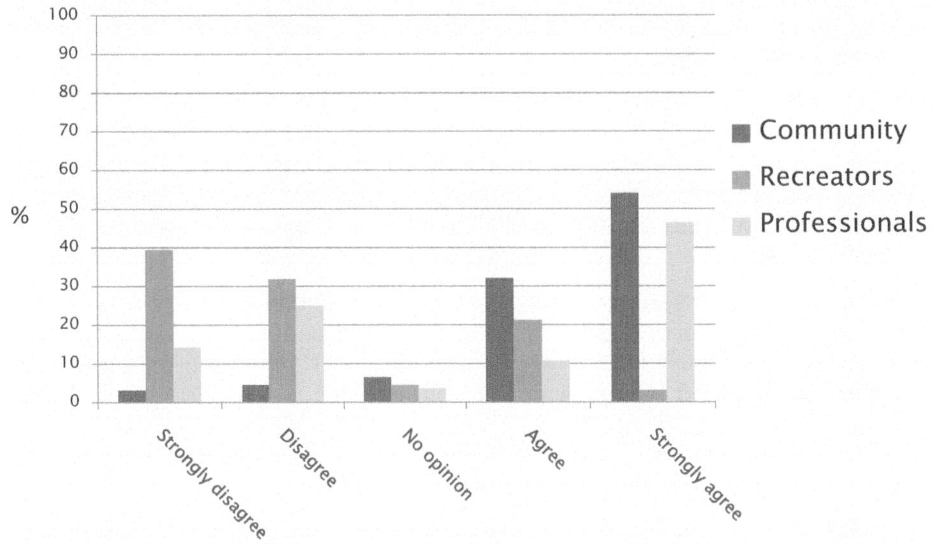

Figure A4.4: Mean Ratings of Agreement with the Policy for Drinking Water Protection

Respondents were asked, unprompted, to say why they thought this way. The main responses follow in Table A4.10 for each sample.

Table A4.10. Unprompted Reasons for Rating Response of Agreement or Disagreement with Policy

Reason	Community % N=604	Recreators % N=66	Professionals % N=28
Must protect drinking water	46.2	16.7	14.3
Stop contamination of drinking water	32.1	10.6	21.4
Cost too much for additional treatment	21.5	0.0	10.7
Need safe water to drink	16.7	9.1	17.9
There are other places to go to recreate	10.6	3.0	17.9
People are irresponsible	7.0	4.5	3.6
It's working/sensible/keep good water quality	2.6	0.0	10.7
Don't want more chemicals/less treatment is better	2.2	0.0	14.3
Prevention is better than the cure	0.5	0.0	21.4
Good for the environment/protects wildlife	0.7	0.0	10.7
No swimming in drinking water	0.8	0.0	3.6
I don't believe recreation pollutes	1.8	24.2	3.6
Some more activities could be allowed (e.g. fishing/swimming)	1.5	3.0	14.3
Recreation is allowed in catchments in other parts of Australia/of the world	0.8	16.7	0.0
Recreation is healthy	1.2	4.5	3.6
Not enough places for recreation	1.0	10.6	0.0
Problem is water birds/(feral) animals	1.0	1.5	0.0
Not policed	1.2	1.5	17.9
Most people are responsible	0.0	12.1	0.0
Should have freedom to roam	0.0	3.0	0.0
Blanket ban is wrong/not balanced	0.0	6.1	0.0
Not based on science	0.0	3.0	0.0
Cost will increase anyway	0.3	0.0	3.6
Other	12.3	9.1	35.7

Respondents were then asked how much they trusted five organisations or groups to provide reliable information on catchment protection and drinking water quality. They were asked to rate their trust on a five point scale from 1 being *no trust at all* to 5 being *complete trust*. The details and mean trust rating of each group and for each sample are shown in the table below.

Table A4.11. Ratings of Trust in Groups to Provide Reliable Information on Catchment Protection and Drinking Water Quality for the Three Samples

Group & Sample	**1 No trust at all %**	**2 %**	**3 Neither %**	**4 %**	**5 Complete trust %**	**Mean Rating**
Water Corporation						
Community (N=604)	6.0	9.6	27.5	38.7	18.2	3.54
Recreators (N=66)	24.2	21.2	27.3	15.2	12.1	2.70
Professionals (N=28)	0.0	14.3	3.6	39.3	42.9	4.11
Department of Health						
Community (N=604)	3.5	6.6	26.2	41.6	22.2	3.72
Recreators (N=66)	9.1	10.6	36.4	31.8	12.1	3.27
Professionals (N=28)	3.6	7.1	3.6	42.9	42.9	4.14
Department of Water						
Community (N=604)	6.3	8.8	29.0	37.06	18.4	3.53
Recreators (N=66)	22.7	15.2	37.9	13.6	10.6	2.74
Professionals (N=28)	7.1	10.7	7.1	39.3	35.7	3.86
Scientists						
Community (N=604)	2.5	5.6	30.5	40.1	21.4	3.72
Recreators (N=66)	7.6	9.1	31.8	31.8	19.7	3.47
Professionals (N=28)	3.6	0.0	17.9	64.3	14.3	3.86
Organised Recreation Groups						
Community (N=604)	11.4	25.0	40.9	17.9	4.8	2.80
Recreators (N=66)	0.0	10.6	40.9	37.9	10.6	3.48
Professionals (N=28)	17.9	50.0	14.3	17.9	0.0	2.32

Oneway analyses of variance to detect statistically significant differences between groups resulted in the following differences ($p<.01$).

- Professionals were significantly more trusting of the Water Corporation than was the community who in turn was significantly more trusting than the recreators.
- Professionals were significantly more trusting of the Department of Health than were recreators.
- Professionals and the community were significantly more trusting of the Department of Water than were recreators.
- Professionals and the community were significantly less trusting of organised recreation groups than were recreators.
- There were no differences in ratings of trust of scientists between the three groups.

Of particular interest were the quite high proportions of respondents in the community and recreator samples who were neither trusting nor distrustful of the groups. Just over one-quarter of community and recreator respondents (27%) were unsure about trust in the Water Corporation to provide reliable information about catchment protection and drinking water quality. Similar proportions of the community respondents were unsure about trust in the Departments of Health and Water, but over a third of recreator respondents neither trusted nor distrusted the Departments of Health (36%) and Water (38%).

There was similar uncertainty about trusting scientists by these two samples (31%), but the "neither" response rose to 40% for both samples in response to trusting organised recreation groups. While this might be expected of the community sample, it is of interest for the recreator sample. Although more respondents in this sample trusted this group than did the other two samples, and none rated *no trust at all*, the uncertainty about "trusting each other" is worthy of note.

The professional sample were more decided about all their trust judgements than the other two samples with only small proportions opting for the "neither" response for any of the groups.

Figure A4.5 below shows the mean trust ratings for each group by each sample.

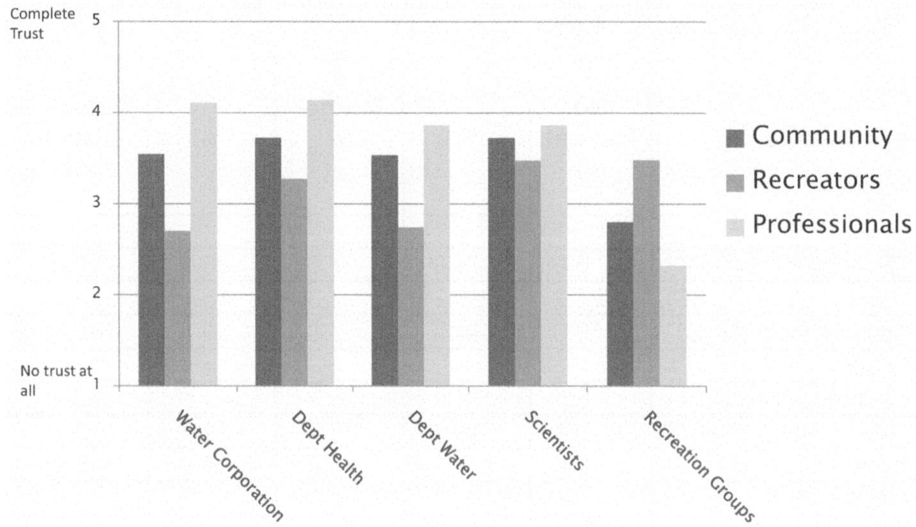

Figure A4.5: Mean Ratings of Trust in Groups to Provide Reliable Information on Catchment Protection and Drinking Water Quality for the Three Samples

ATTITUDES ASSOCIATED WITH CATCHMENT PROTECTION AND RECREATION

Respondents were asked to rate, on a five point scale, their agreement with twenty-eight statements that were designed to measure a number of attitudinal dimensions which may affect decision making about drinking water catchment protection and recreation. The order of the statements was rotated to avoid order bias.

The following table details the ratings of the individual statements for each sample. They are provided here for interest, and discussion of outcomes are provided in the Section, Data Reduction, which describes the formation of attitudinal scales. However, the following will assist the reader in interpretation of the frequencies and means.

- **Bold print** indicates where there is a split opinion across the sample.
- ***Bold italics*** indicates a high proportion of ambivalent respondents in the sample to that statement (i.e. one-quarter or more rated it at 3/neither)
- Statistically significant differences (p<.01) between samples are denoted with asterisks, with the number of asterisks indicating similar groups. *However, please note that many of the differences are only in terms of degree rather than opposing views.*

Table A4.12. Ratings of Agreement with Individual Attitudinal Statements

Statement & Sample	1 Strongly disagree	2 Disagree	3 Neither	4 Agree	5 Strongly agree	Mean Rating
I really haven't thought much about managing catchments to protect my drinking water before today						
* Community (N=604)	**10.3**	**30.3**	**14.9**	**31.8**	**12.7**	**3.06**
** Recreators (N=66)	42.4	48.5	6.1	3.0	0.0	1.70
** Professionals (N=28)	71.4	21.4	0.0	0.0	7.1	1.50
Being able to recreate in the drinking water catchments is important to me.						
* Community (N=604)	33.6	35.6	10.9	15.1	4.8	2.22
** Recreators (N=66)	1.5	3.0	9.1	28.8	57.6	4.38
* Professionals (N=28)	46.4	14.3	7.1	32.1	0.0	2.25
Prevention is better than the cure in protecting our drinking water quality						
* Community (N=604)	0.7	1.7	6.1	39.7	51.8	4.40
** Recreators (N=66)	1.5	12.1	***27.3***	45.5	13.6	3.58
* Professionals (N=28)	0.0	3.6	7.1	14.3	75.0	4.61
I have complete trust in the authorities to manage our water responsibly						
* Community (N=604)	5.8	16.9	***26.3***	42.4	8.6	3.31
** Recreators (N=66)	13.6	34.8	19.7	28.8	3.0	2.73
* Professionals (N=28)	3.6	17.9	3.6	53.6	21.4	3.71
Dams currently used for recreation should not be taken away for drinking water in the future						
* Community (N=604)	**10.9**	**31.6**	**19.5**	**25.3**	**12.6**	**2.97**
** Recreators (N=66)	1.5	15.2	18.2	33.3	31.8	3.79
* Professionals (N=28)	3.6	39.3	10.7	32.1	14.3	3.14
Human concerns should come before the natural environment						
Community (N=604)	19.0	36.1	***26.2***	14.2	4.5	2.49
Recreators (N=66)	7.6	45.5	***33.3***	12.1	1.5	2.55
Professionals (N=28)	39.3	28.6	***25.0***	3.6	3.6	2.04
Compared with other things, protecting the quality of drinking water is not a major issue in my life						
Community (N=604)	24.0	41.4	16.7	15.4	2.5	2.31
* Recreators (N=66)	13.6	38.4	22.7	27.3	1.5	2.68
** Professionals (N=28)	42.9	35.7	14.3	3.6	3.6	1.89
Technology can provide whatever treatment is needed to produce safe drinking water.						
* Community (N=604)	4.0	14.9	***25.2***	42.1	13.9	3.47
** Recreators (N=66)	4.5	7.6	9.1	34.8	43.9	4.06
* Professionals (N=28)	17.9	17.9	10.7	46.4	7.1	3.07
When it comes to providing drinking water, the Water Corporation is committed to an unbiased process of decision making						
* Community (N=604)	5.0	16.9	***35.9***	36.8	5.5	3.21
** Recreators (N=66)	30.3	39.4	16.7	9.1	4.5	2.18
* Professionals (N=28)	14.3	35.7	14.3	17.9	17.9	2.89
It is safer to drink water from a clean environment than to rely on artificial treatment						
* Community (N=604)	2.3	12.7	21.5	35.8	27.6	3.74
** Recreators (N=66)	9.1	47.0	13.6	22.7	7.6	2.73
* Professionals (N=28)	0.0	14.3	14.3	25.0	46.4	4.04

Table A4.12 cont. Ratings of Agreement with Individual Attitudinal Statements

Statement & Sample	1 Strongly disagree	2 Disagree	3 Neither	4 Agree	5 Strongly agree	Mean Rating
People should have the right to recreate anywhere that is public land						
** Community (N=604)	25.8	43.9	14.7	12.3	3.3	2.23
*** Recreators (N=66)	9.1	39.4	10.6	33.3	7.6	2.91
* Professionals (N=28)	57.1	39.3	0.0	0.0	3.6	1.54
I have very definite thoughts about protecting drinking water						
Community (N=604)	1.3	8.8	21.7	41.9	26.3	3.83
Recreators (N=66)	0.0	9.1	13.6	57.6	19.7	3.88
Professionals (N=28)	0.0	0.0	14.3	39.3	46.4	4.32
I insist on being able to recreate anywhere I choose						
** Community (N=604)	27.8	50.5	11.9	7.3	2.5	2.06
*** Recreators (N=66)	6.1	54.5	16.7	18.2	4.5	2.61
* Professionals (N=28)	60.7	39.3	0.0	0.0	0.0	1.39
I value having areas that exclude human activity and can provide protection for endangered species						
* Community (N=604)	0.8	2.2	4.8	41.6	50.7	4.39
** Recreators (N=66)	4.5	9.1	12.1	57.6	16.7	3.73
* Professionals (N=28)	3.6	3.6	0.0	32.6	60.7	4.43
Protecting our drinking water now is protecting our children's future						
* Community (N=604)	0.7	1.2	4.6	37.1	56.5	4.48
** Recreators (N=66)	9.1	16.7	16.7	42.4	15.2	3.38
* Professionals (N=28)	0.0	0.0	7.1	35.7	57.1	4.50
I have complete trust in the authorities to provide me with safe tap water to drink						
Community (N=604)	3.3	12.9	19.9	49.3	14.6	3.59
* Recreators (N=66)	4.5	19.7	13.6	51.5	10.6	3.44
** Professionals (N=28)	0.0	10.7	10.7	46.4	32.1	4.00
When it comes to catchment protection, we should be more concerned about mining and forestry than recreation						
* Community (N=604)	6.1	15.1	22.4	38.2	18.2	3.47
* Recreators (N=66)	1.5	18.2	13.6	37.9	28.8	3.74
** Professionals (N=28)	14.3	42.9	17.9	21.4	3.6	2.57
The Water Corporation is an important member of the Perth Community						
* Community (N=604)	2.2	3.0	16.2	54.6	24.0	3.95
** Recreators (N=66)	7.6	10.6	13.6	59.1	9.1	3.52
* Professionals (N=28)	0.0	7.1	7.1	46.4	39.3	4.18
Recreation poses very little risk to drinking water quality						
** Community (N=604)	17.5	45.2	20.7	13.7	2.8	2.39
*** Recreators (N=66)	0.0	24.2	24.2	39.4	12.1	3.39
* Professionals (N=28)	39.3	39.3	14.3	7.1	0.0	1.89
I believe the authorities should go to any lengths to protect drinking water catchments						
*** Community (N=604)	1.8	9.4	13.4	41.1	34.3	3.97
* Recreators (N=66)	18.2	51.5	10.6	13.6	6.1	2.38
** Professionals (N=28)	7.1	7.1	***42.9***	17.9	25.0	3.46

Table A4.12 cont. Ratings of Agreement with Individual Attitudinal Statements

Statement & Sample	1 Strongly disagree	2 Disagree	3 Neither	4 Agree	5 Strongly agree	Mean Rating
Protecting drinking water quality is not a personal interest of mine						
** Community (N=604)	25.2	43.2	13.9	13.4	4.3	2.28
Recreators (N=66)	22.7	53.0	18.2	3.0	3.0	2.11
* Professionals (N=28)	57.1	28.6	7.1	3.6	3.6	1.68
I trust the authorities to keep up with the latest water treatment science						
Community (N=604)	2.2	5.5	16.7	57.1	18.5	3.84
* Recreators (N=66)	1.5	21.2	12.1	48.5	16.7	3.58
** Professionals (N=28)	0.0	7.1	10.7	50.0	32.1	4.07
It's too risky to allow people to recreate close to, or on the drinking water dams						
* Community (N=604)	3.8	8.4	10.9	40.4	36.4	3.97
** Recreators (N=66)	25.8	47.0	12.1	10.6	4.5	2.21
* Professionals (N=28)	0.0	7.1	***32.1***	21.4	39.3	3.93
I don't believe the Water Corporation is part of the Perth Community						
* Community (N=604)	15.1	46.9	***26.2***	9.8	2.2	2.37
* Recreators (N=66)	7.6	48.5	19.7	15.2	9.1	2.70
** Professionals (N=28)	39.3	42.9	10.7	7.1	0.0	1.86
People have the right to have good quality drinking water						
* Community (N=604)	0.0	0.3	0.2	26.0	73.5	4.73
** Recreators (N=66)	0.0	1.5	3.0	48.5	47.0	4.41
* Professionals (N=28)	0.0	0.0	0.0	32.1	67.9	4.68
Keeping areas of untouched bushland is important for future generations						
* Community (N=604)	0.2	1.5	3.3	33.1	61.9	4.55
** Recreators (N=66)	3.0	9.1	6.1	53.0	28.8	3.95
* Professionals (N=28)	0.0	7.1	7.1	25.0	60.7	4.39
The Water Corporation makes an effort to treat everyone fairly						
* Community (N=604)	3.1	9.8	***30.3***	46.5	10.3	3.51
** Recreators (N=66)	19.7	28.8	24.2	24.2	3.0	2.62
* Professionals (N=28)	3.6	21.4	7.1	42.9	25.0	3.64
The use of dams for drinking water should always take priority over recreation						
* Community (N=604)	0.5	5.5	8.9	39.4	45.7	4.24
** Recreators (N=66)	12.1	36.4	24.2	22.7	4.5	2.71
* Professionals (N=28)	0.0	21.4	10.7	21.4	46.4	3.93

SYSTEM RISK

Respondents were asked the following three part question to measure perceptions of system risk associated with allowing recreation in drinking water catchments. All questions were answered using five point scales.

If recreation was allowed in the 2km restricted zone in drinking water catchments, as well as in and on the water, and additional treatment plants were installed:

- *How likely do you think it is that something could go wrong with the quality of your drinking water?*
- *How serious do you think it would be if something went wrong with the quality of you drinking water?*
- *What level of control do you think the government authorities would have to stop something going wrong with the quality of your drinking water?*

The following figures show the mean responses of the three samples to the three questions.

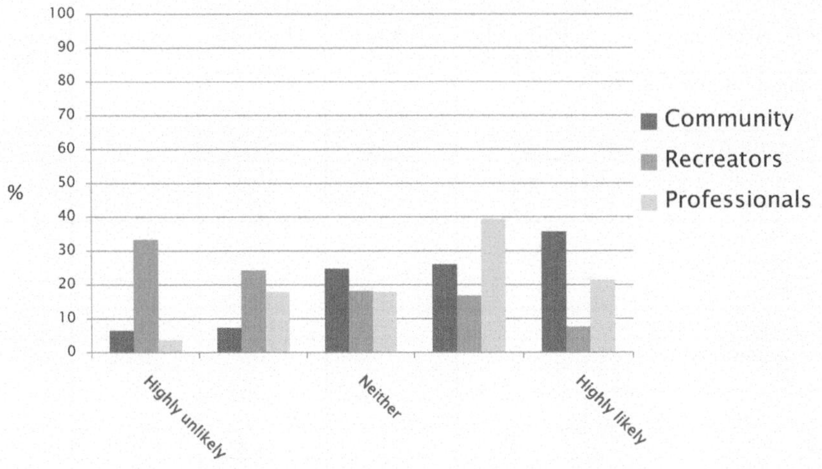

Figure A4.6: Mean Ratings of Likelihood that Something could go Wrong with the Quality of Drinking Water

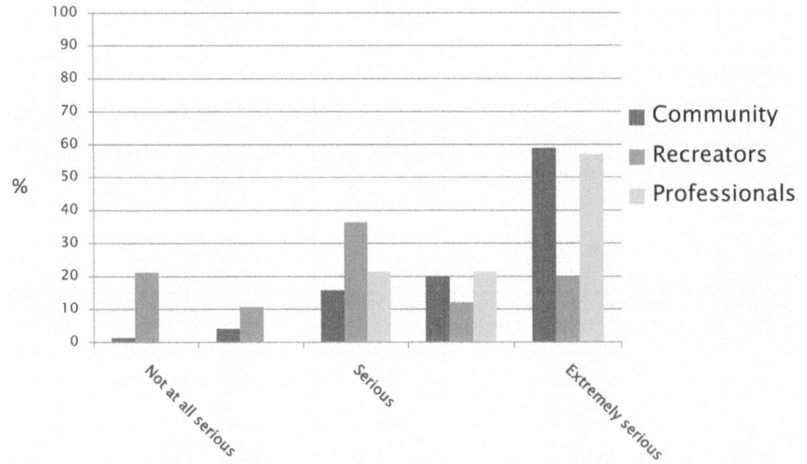

Figure A4.7. Mean Ratings of Seriousness if Something went Wrong with the Quality of Drinking Water

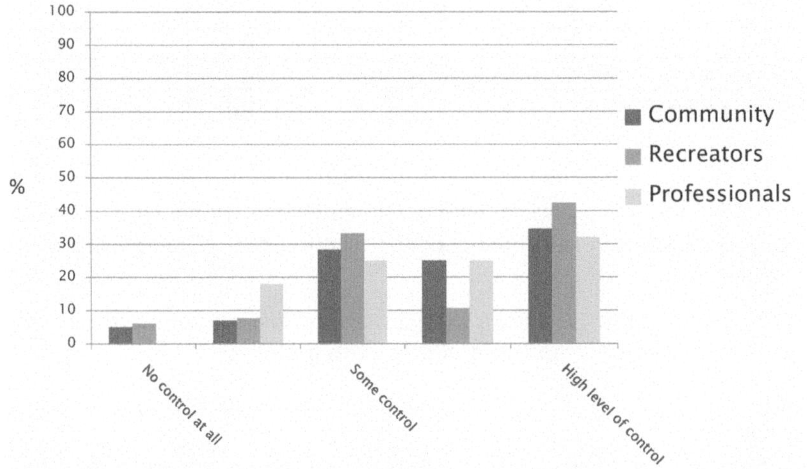

Figure A4.8. Mean Ratings of Perceived Level of Control the Government Authorities would have to stop Something going Wrong with the Quality of Drinking Water

Responses of the community sample indicated that they thought it likely that something would go wrong, that it would be serious if it did, but that the Government authorities would have a level of control to stop it going wrong.

Similar responses were provided by the professionals, while the recreators considered a water quality incident to be significantly less likely and significantly less serious ($p<.01$) than both the community and the professional samples. However, there was no significant difference in perceptions of levels of control between the three samples.

POLICY PREFERENCES

Respondents were provided with the following information before being asked which of six policy options they thought were acceptable and which were unacceptable. They were then asked to nominate which option was *most acceptable* out of all those that were deemed acceptable, and which was *most unacceptable* out of all those that were deemed unacceptable. The policy options were presented in order of increasing risk associated with increased activities and increasing contact with water. This assisted respondents in their consideration of options through the telephone interview. Interviewers were also provided with prompts to assist people in their assessments of *most acceptable* and *most unacceptable*.

> *As we said previously, the current policy allows no human contact with the water in the dams, and no public access within 2kms of the water.*
>
> *If any of this policy changed, additional treatment would be needed to provide safe drinking water. This would increase the cost of water to householders by 15%.* ***This same level of treatment would be needed whether a <u>small number</u> of extra activities, or <u>a large number</u> of extra activities were allowed.***
>
> *Different activities pollute the water in different ways, for example, bacteria and pathogens from human activities, hydro-carbons from motorised vehicles. Some activities are more polluting than others.* ***More activities mean more pollutants and these put increasingly greater pressure on the treatment processes. The more the activities, the more the overall risk of something going wrong increases.***

The following table shows the percentages of each sample that found each policy most acceptable, acceptable, unacceptable and most unacceptable.

Table A4.13. Acceptability of Catchment Management Policies

Policy Acceptability	Community %	Recreators %	Professionals %
Keep the current policy with no additional cost and no increased treatment risk.	**N=595**	**N=63**	**N=28**
Most unacceptable	2.9	57.1	7.1
Unacceptable	8.4	17.5	17.9
Acceptable	21.3	15.9	7.1
Most acceptable	67.4	9.5	67.9
Allow off-road activities (such as trail bikes and 4 WD) in the outer catchment, but still no activities in the 2km restricted zone and no contact with water, at added cost and greater risk to your drinking water quality.	**N=593**	**N=63**	**N=28**
Most unacceptable	6.6	6.3	3.6
Unacceptable	55.1	38.1	78.6
Acceptable	31.9	34.9	17.9
Most acceptable	6.4	20.6	0.0
Allow passive activities in the 2km restricted zone (such as walking and picnicking), but no contact with water, at added cost and greater risk to your drinking water quality.	**N=597**	**N=65**	**N=28**
Most unacceptable	1.0	1.5	0.0
Unacceptable	53.8	20.0	60.7
Acceptable	31.0	43.1	17.9
Most acceptable	14.2	35.4	21.4

Table A4.13 cont'd. Acceptability of Catchment Management Policies

Policy Acceptability	Community %	Recreators %	Professionals %
Allow camping and off-road activities in addition to the passive activities in the 2km restricted zone, but no contact with water, at added cost and greater risk to your drinking water quality.	**N=598**	**N=63**	**N=28**
Most unacceptable	3.2	0.0	0.0
Unacceptable	77.1	68.3	89.3
Acceptable	15.6	25.4	3.6
Most acceptable	4.2	6.3	7.1
Allow passive activities, camping and off-road activities <u>throughout the whole catchment</u>, and fishing, swimming, canoeing and sailing in the water, but no motorised boats, at added cost and greater risk to your drinking water quality.	**N=598**	**N=61**	**N=28**
Most unacceptable	3.8	0.0	10.7
Unacceptable	83.4	59.0	85.7
Acceptable	8.4	14.8	3.6
Most acceptable	4.3	26.2	0.0
Allow passive activities, camping and off-road activities <u>throughout the whole catchment</u>, and fishing, swimming, canoeing, sailing and motorised boats in the water, at added cost and greater risk to your drinking water quality.	**N=598**	**N=62**	**N=28**
Most unacceptable	81.1	37.1	78.6
Unacceptable	15.2	50.0	17.9
Acceptable	2.2	6.5	0.0
Most acceptable	1.5	6.5	3.6

Clearly the current policy was the most acceptable to the community sample with 89% considering it to be *acceptable* and of those two-thirds considered it to be *most acceptable*. However, this also means that one-third of the community sample considered other policy options to be *most acceptable*. The next acceptable policy options to the community were the two following with off-road activities allowed in the outer catchment and no activities in the 2km zone (38.3% considered it *acceptable,* of which 6.4% considered it to be *most acceptable*), and passive activities only allowed in the 2km zone (45.2% considered it *acceptable,* of which 14.2% considered it to be *most acceptable*).

The most unacceptable policy option to the community was that with unrestricted activities throughout the catchment and water with 81.1% considering it to be *most unacceptable*.

The current policy was the most unacceptable to the recreator sample, with a total of 75% finding it *unacceptable* of which 57% considered it to be *most unacceptable*. However, while the sixth and totally unrestricted policy was found to be *unacceptable* by 87% of recreators, only 37% thought it to be *most unacceptable*. The most acceptable policy to the recreators was the third that allowed greater access to the 2km restricted zone with 79% finding it *acceptable* of which 35% found it to be *most acceptable*. The second policy allowing greater access to the outer catchment showed 55% of recreators found it to be *acceptable* of which 21% found it to be *most acceptable*. However, while 26% considered the policy that allowed access to water by non-motorised activities to be *most acceptable*, 59% of recreators found it to be *unacceptable*.

The professionals' considerations were not unlike those of the community with two-thirds finding the current policy to be *most acceptable*, but one-third considering other options to be *most acceptable*. Of these, the third option of greater activity in the 2km restricted zone was the preferred, with 21% of professionals considering this to be *most acceptable*.

The following statistically significant differences ($p<.01$) between samples emerged.

- Recreators considered the current policy to be less acceptable than did the community and professionals.
- Recreators considered the third, fifth and sixth policy options to be more acceptable than did both the community and professionals.
- Recreators considered the second policy option to be significantly more acceptable than did the professionals, but not different from the community. The community was also not significantly different from the professionals.

- There were no significant differences in considerations of acceptability of the fourth policy option by the three samples (unrestricted land activities but restricted water contact).

Figures A4.9 to A4.12 summarise the acceptability and unacceptability of the policy options for each sample.

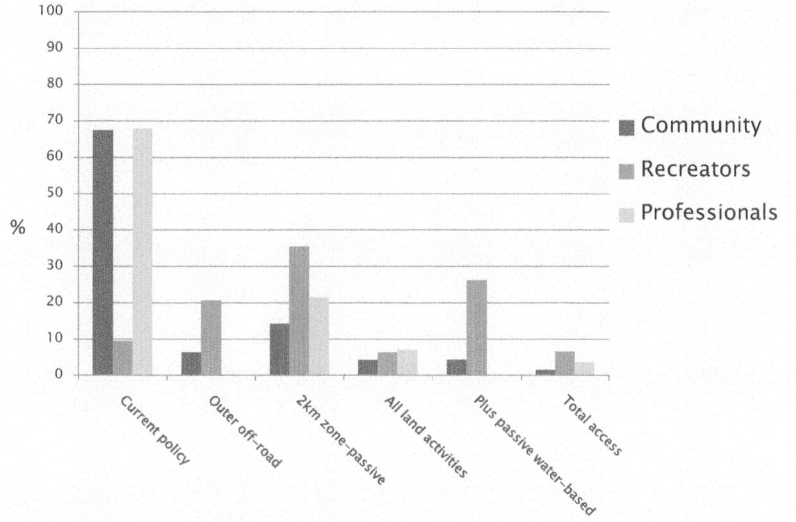

Figure A4.9. Most Acceptable Policy Options for Each Sample

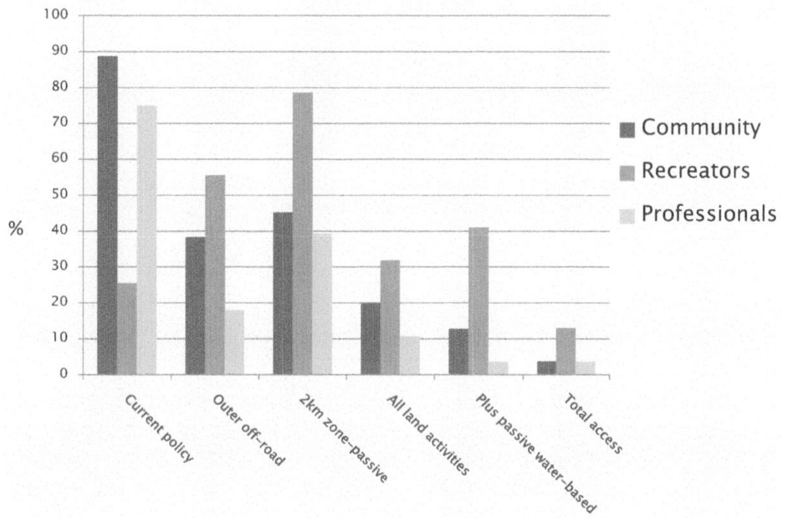

Figure A4.10. Most Acceptable and Acceptable (Combined) Policy Options for Each Sample

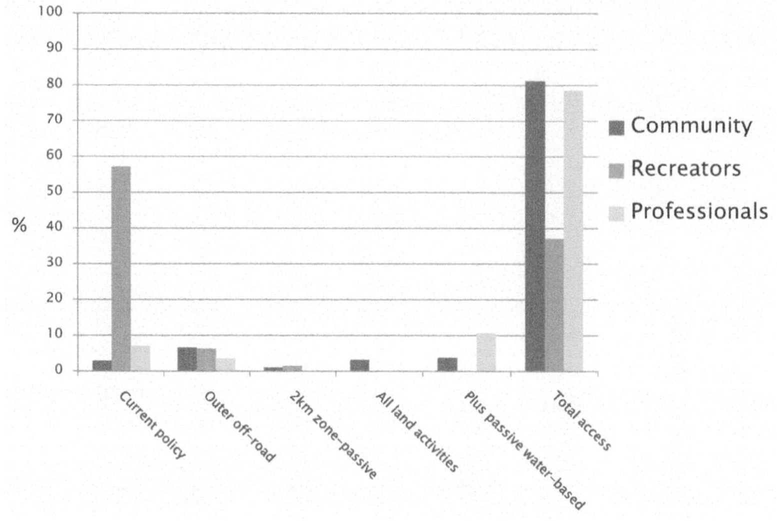

Figure A4.11. Most Unacceptable Policy Options for Each Sample

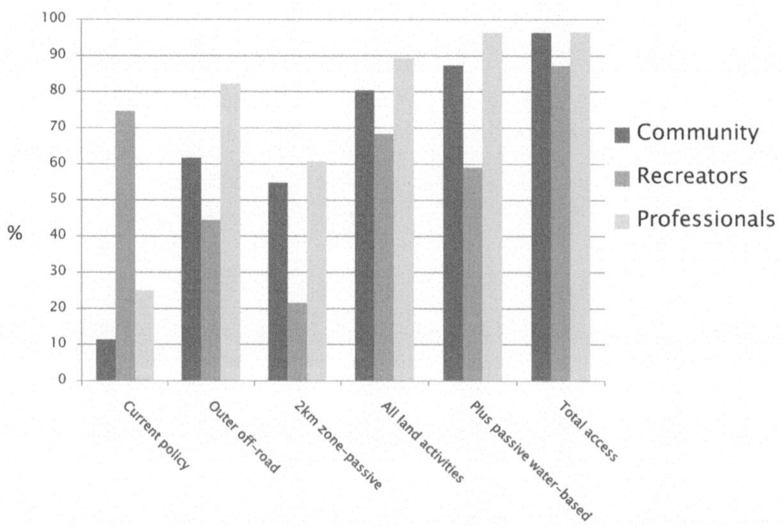

Figure A4.12. Most Unacceptable and Unacceptable (Combined) Policy Options for Each Sample

Respondents were then asked if any of their policy preferences would change if the drinking water catchment was in a rural location and it was the only water body available to the community for recreation. Sixty-nine percent of the community sample and 74% of the recreators said that their preferences would not change, while 23% of the community and 21% of the recreators said that they would. Eight percent of the community and 5% of the recreators were not sure. However, 54% of the professionals said their preferences would not change, while 25% were not sure.

Those who answered *yes* or *not sure* (N=185 community; 17 recreators; 13 professionals[16]) were asked how their preferences would change. Twenty-six percent of community, 24% of recreators and 23% of professionals said that they'd allow *most activities in the water (especially swimming) but not allow motorised boats*. Twenty-one percent of community and 24% of recreators said they'd allow *passive activities next to the water, but no contact with the water*. Eight percent of the community and 12% of recreators said they'd *allow everything*, while 5% of the community said there'd be *no choice and they'd have to pay for extra treatment*. Twenty-three percent of the professionals said there could be *limited and well managed and controlled activities*.

---

[16] Please note percentages in the following sentences for recreators and professionals involve very small numbers.

Respondents were finally asked who should pay for the additional treatment that would be required if the policy was relaxed to allow greater recreation in the drinking water catchments. Community and recreator respondents were asked unprompted, while the professionals were provided with the options.

Table A4.14. Who should pay for Additional Treatment

Who pays	Community % N=604	Recreators % N=66	Professionals % N=27
Recreators	15.9	3.0	40.7
Users	32.3	19.7	14.8
Everyone	24.3	59.1	40.7
Government	23.5	7.6	3.7
Other	4.0	10.6 [1]	0.0

[1] All seven of the recreators who provided the "other" response said they *didn't believe extra treatment would be required.*

About a quarter of the community sample (24%), 59% of recreators and 41% of professionals nominated *everyone* and about a further quarter of the community (24%), 8% of recreators and 4% of professionals nominated the *Government*.

Almost half the community sample (48%), 23% of recreators and 56% of professionals nominated either *recreators* or *users* and these respondents (N=291community and 15 each of recreators and professionals) were asked a further question: *There is considerable evidence that recreation increases the health and well-being of society and so may reduce the government cost of health services. Do you still think that recreators/users should pay?* Ninety percent of the community, 93% of recreators and 94% of professionals answered "yes".

DEMOGRAPHICS

Table A4.15. Membership of Organised Recreation Group

	Community % N=604	Recreators % N=66	Professionals % N=28
Yes	20.7	100.0	32.1

Table A4.16. Requirement for Greater Access to Drinking Water Catchments

Greater Access	Community % N=604	Recreators % N=66	Professionals % N=28
Yes	11.1	86.4	10.7
Maybe	3.6	1.5	10.7
No	85.3	12.1	78.6

Table A4.17. Age

Age Category	Community % N=604	Recreators % N=66	Professionals % N=28
18-24 years	8.1	0.0	0.0
24 – 39 years	41.7	16.7	28.6
40 – 55 years	22.0	40.9	53.6
56 – 65 years	13.7	25.8	14.3
66 – 75 years	10.6	13.6	3.6
More than 75 years	3.8	3.0	0.0

Table A4.18. Family Unit

Unit Description	Community % N=604	Recreators % N=66	Professionals % N=28
Single adult less than 65 years	8.4	12.1	3.6
Single adult more than 65 years	6.0	3.0	0.0
Two adults – older person less than 65 years	18.0	31.8	42.9
Two adults – older person more than 65 years	10.3	10.6	3.6
Single adult – eldest child less than 18 years	1.8	1.5	0.0
Single adult – eldest child more than 18 years	2.2	0.0	0.0
Two adults – eldest child less than 18 years	32.8	33.3	35.7
Two adults – eldest child more than 18 years	11.1	7.6	14.3
More than two adults – no children	5.0	0.0	0.0
More than two adults – eldest child less than 18 years	3.0	0.0	0.0
More than two adults – eldest child more than 18 years	1.5	0.0	0.0

Table A4.19. Region of Birth

Region	Community % N=604	Recreators % N=64	Professionals % N=28
Australia	69.9	59.1	57.1
New Zealand/Pacific	2.8	1.5	0.0
United Kingdom/Ireland	14.9	31.8	28.6
Western Europe/Scandinavia	2.0	0.0	7.1
Eastern Europe/Russia	0.7	0.0	3.6
USA/Canada	0.8	1.5	0.0
Asia/India	4.3	1.5	3.6
Middle East	0.7	0.0	0.0
Central/South America	0.7	0.0	0.0
Africa	3.3	1.5	0.0

Table A4.20. Time of Residence in WA

Years	Community % N=604	Recreators % N=65	Professionals % N=28
Less than 5 years	6.8	0.0	0.0
5 to 10 years	6.8	4.5	14.3
11 to 20 years	9.8	10.6	7.1
More than 20 years	76.7	83.3	78.6

DATA REDUCTION

A number of scales were formed by combining rating responses to several statements or questions and their reliability tested for use in further analyses, including multi-variate analysis[17]. An indicator of reliability is internal consistency, or how the items in a scale 'hang together'. The most commonly used indicator of internal consistency is Cronbach's alpha coefficient. This provides an indication of the average correlation among all of the items on a particular scale. A scale's coefficient will ideally be above 0.7, however scales with fewer than ten items commonly receive quite low coefficients on this measure (Pallant, 2004, p. 90). In these cases, Cronbach's alpha above 0.5 can be considered acceptable. The following outlines the scales that resulted in sufficient reliability for use in further analyses. Some items were recoded to ensure the same directionality of measurement and these are noted below.

[17] Multi-item scales provide variable measures that are more reliable for on-going analyses than are responses to single item attitudinal statements (e.g. as shown in Tables 3.3 and 3.11) or to individual questions as in Section 3.4.

Recreational Enjoyment

The following four statements about ability to enjoy recreation in places of choice formed a reliable scale: Cronbach's α = .66.

- *There are only a few places I'm allowed to go to enjoy my recreational activities (recoded)*
- *More and more people are coming to my favourite places to recreate(recoded)*
- *My recreational activity is being banned in more and more places (recoded)*
- *My recreational enjoyment is often lessened by other activities in the same place(recoded)*

Possible scores on the *recreational enjoyment* scale ranged from a minimum of 1 (low enjoyment) through to a maximum of 5 (high enjoyment).

The mean score for the community sample of 3.3 indicated general enjoyment of their recreational activities. However, 29% of the sample scored less than 3, signifying some barriers to recreational enjoyment by a notable proportion of the community sample.

The mean score for the recreator sample of 2.2 indicated a lack of enjoyment of their recreational activities with 80% (N=52) of the sample scoring less than 3. This signifies considerable issues associated with the recreators' abilities to enjoy their activities of choice, and given the diversity of activities represented in the sample this is of note.

The mean score for the professional sample of 3.5 indicated general enjoyment of their recreational activities with 21% of the sample scoring less than 3.

Statistically significant differences ($p<.01$) between samples showed the recreators to be enjoying their activities less than the community and professional samples.

Trust in Authorities

The following five statements about trust in authorities in regard to drinking water quality formed a reliable scale: Cronbach's α = .74.

- *I have complete trust in the authorities to provide me with safe tap water to drink*
- *I have complete trust in the authorities to manage our water responsibly*
- *I don't trust the authorities to keep up with the latest water treatment science (recoded)*
- *The Water Corporation is an important member of the Perth Community*
- *I don't believe the Water Corporation is part of the Perth Community (recoded)*

Possible scores on the *trust in authorities* scale ranged from a minimum of 1 (low trust) through to a maximum of 5 (high trust).

The mean score for the community sample of 3.7 indicates general trust in the authorities to provide good quality drinking water with 66% scoring more than 3. Only 5% scored less than 3.

The mean score for the recreator sample of 3.3 indicates some uncertainty in trust in the authorities with 24% scoring more than 3 and 24% scoring less than 3.

The mean score for the professional sample was 4.0 with 57% scoring more than 3 and 11% less than 3.

Statistically significant differences ($p<.01$) between samples showed the recreators to be least trusting of authorities, with the community being more trusting than recreators but less trusting than the professionals.

Catchment Protection

The following four statements about catchment protection and allowing recreation in drinking water catchments formed a reliable scale: Cronbach's α = .72.

- *I believe the authorities should go to any lengths to protect drinking water catchments*
- *I insist on being able to recreate anywhere I choose (recoded)*
- *Being able to recreate in the drinking water catchments is important to me (recoded)*
- *The use of dams for drinking water should always take priority over recreation*

Possible scores on the *catchment protection* scale ranged from a minimum of 1 (anti-protection) through to a maximum of 5 (pro-protection).

The mean score for the community sample of 4.0 indicates agreement with protecting drinking water catchments. Interestingly no-one scored 1 on the scale and 80% scored more than 3.

The mean score for the recreator sample of 2.5 indicates favouring recreation in drinking water catchments with 70% of this sample scoring less than 3.

The mean score for the professional sample of 3.9 indicates a similar stance to that of the community with 57% scoring more than 3.

Statistically significant differences ($p<.01$) between samples showed the recreators less in favour of protection of drinking water catchments than either the community or the professionals.

Precautionary Ethics

The following seven precautionary ethical statements (i.e. human and environmental caution in both the present and the future) formed a reliable scale: Cronbach's $\alpha = .70$.

- *Protecting our drinking water now is protecting our children's future*
- *Ensuring areas of untouched bush land is important for future generations*
- *I value having areas that exclude human activity and can provide protection for endangered species*
- *Human concerns come before the natural environment (recoded)*
- *People have the right to have good quality drinking water*
- *Prevention is better than the cure in protecting our drinking water quality*
- *People should have the right to recreate anywhere that is public land (recoded)*

Possible scores on the *precautionary ethics* scale ranged from a minimum of 1 (less conservative) through to a maximum of 5 (more conservative).

The mean score for the community sample of 4.3 indicates highly conservative precautionary ethics. No-one scored 1 on the scale and 93% scored more than 3.

The mean score for the recreator sample of 3.7 indicates lesser conservatism with one-third scoring more than 3 and 12% scoring less than 3. Half the sample, though, were uncertain.

The mean score for the professional sample of 4.4 shows similar conservatism to the community. Seventy-nine percent of this sample scored more than 3.

Statistically significant differences ($p<.01$) between samples showed the recreators to be less conservative in their precautionary ethics than the community and the professionals, but this difference was a matter of degree and not opposing in ways of thinking.

Fair Procedures

The following two statements in regard to the Water Corporation's decision making processes in providing drinking water formed a reliable scale: Cronbach's $\alpha = .69$.

- *When it comes to providing drinking water, the Water Corporation is committed to an unbiased process of decision making*
- *The Water Corporation makes an effort to treat everyone fairly.*

Possible scores on the *fair procedures* scale ranged from a minimum of 1 (lesser fairness) through to a maximum of 5 (greater fairness).

The mean score for the community sample of 3.4 indicates general considerations of fairness by the community in the Water Corporation to provide fair decision making processes, with about half the sample scoring 4 or 5, and a further quarter scoring 3. Only 10% scored less than 3.

The mean score for the recreator sample of 2.4 indicates quite low levels of trust in the Water Corporation to provide fair decision procedures. Only 9% scored more than 3, and 59% scored less than 3.

The mean score for the professional sample of 3.3 is again similar to that of the community sample, but with a spread of opinion across the scale.

Statistically significant differences ($p<.01$) between samples showed the recreators to be significantly less trusting than the community and the professionals.

Strength of Attitudes

The following four statements showing strength of attitudes about drinking water and its protection formed a reliable scale: Cronbach's $\alpha = .51$.

- *I really haven't thought much about managing catchments to protect my drinking water before today (recoded)*
- *Protecting drinking water quality is not a personal interest of mine (recoded)*
- *Compared with other things, protecting drinking water quality is not a major issue in my life (recoded)*
- *I have very definite thoughts about protecting drinking water*

Possible scores on the *strength of attitudes* scale ranged from a minimum of 1 (weaker attitudes) through to a maximum of 5 (stronger attitudes).

The mean score for the community sample of 2.8 indicates uncertain attitudes about drinking water and its protection. No-one scored 5 and only 10% of the sample scored 4. Almost two-thirds of the sample, 61%, scored 3.

The mean score for the recreator sample of 3.1 indicates uncertain attitudes about drinking water and its protection similar to that of the community. Only 2% scored more than 3 and 35% scored less than 3.

The mean score for the professional sample of 3.5 indicates moderately strong attitudes with 32% scoring more than 3 and only 14% less than 3.

Statistically significant differences ($p<.01$) between samples showed the professionals to have stronger attitudes about drinking water and its protection than both the community and the recreators.

Catchment Risk

The following three statements about risk associated with catchment management and recreation formed a reliable scale: Cronbach's $\alpha = .62$.

- *Recreation poses very little risk to drinking water quality (recoded)*
- *It's too risky to allow people to recreate close to, or on the drinking water dams*
- *It is safer to drink water from a clean environment than to rely on artificial treatment*

Possible scores on the *catchment risk* scale ranged from a minimum of 1 (risk tolerant) through to a maximum of 5 (risk averse).

The mean score for the community sample of 3.8 indicates a generally risk averse sample with two-thirds scoring more than 3 and about a quarter scoring 3.

The mean score for the recreator sample of 2.5 indicates quite a risk tolerant group, with only 4% scoring more than 3 and 70% scoring less than 3.

The mean score for the professional sample of 4.0 indicates they are similarly risk averse to the community. Fifty-seven percent scored more than 3 with only 11% scoring less than 3.

Statistically significant differences ($p<.01$) between samples showed the recreators to be significantly less concerned about the risk to their drinking water from recreation than both the community and the professionals.

System Risk

A *system risk* score was calculated by summing the ratings of the three questions described previously:

- *perceived likelihood of something going wrong if recreation was allowed in drinking water catchments*
- *perceived seriousness if something went wrong;*
- *perceived government control to stop something going wrong (recoded).*

Possible *system risk* scores ranged from 3 being low risk (likelihood and seriousness) and high control through to 15 being high risk and low control.

The mean score for the community sample was 10.3 which indicated general concern about the risk of a drinking water quality incident if the catchments were opened to recreation. Two-thirds of the sample scored more than 9.

The mean score for the recreator sample of 7.6 indicated considerations of lower risk with control. Fifty percent scored less than 8 and only 20% scored more than 9.

The mean score for the professional sample of 10.2 was again similar to that of the community with 60% scoring more than 9 and only 11% scoring less than 8.

Statistically significant differences (p<.01) between samples again showed the recreators to be less concerned about risk to their drinking water through allowing recreation in the catchments than both the community and the professionals.

Figures A4.13 and A4.14 below depict the mean scores for each sample on each attitudinal scale.

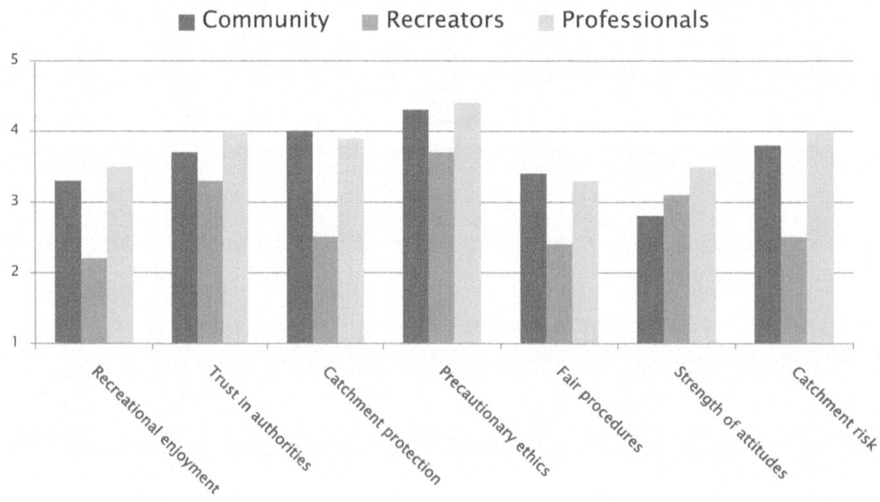

Figure A4.13: Mean Scores on Attitudinal Scales

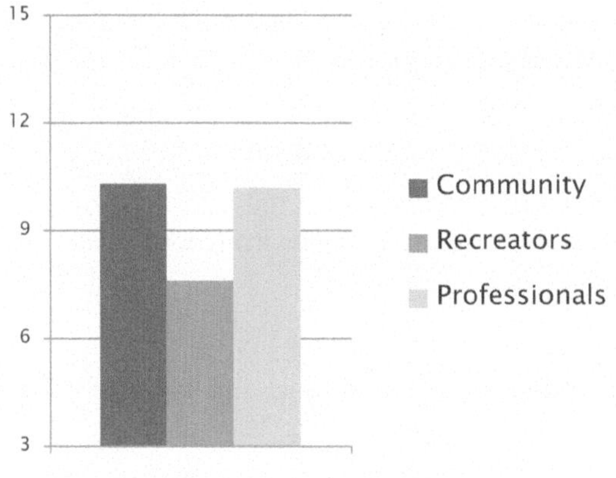

Figure A4.14: Mean Scores on the *System Risk* Attitudinal Scale

MULTI-VARIATE ANALYSES

A Discriminant analysis, with stepwise method, was conducted to determine which variables were the major discriminators of **acceptance of the current policy** of catchment protection, the four groups being most unacceptable, unacceptable, acceptable and most acceptable. All three samples were combined for the analysis (N=686). The independent variables in the analysis were:

- actual *knowledge* score
- *trust* in each of the Water Corporation, Departments of Health and Water, Scientists, and Organised Recreation Groups to provide reliable information about catchment protection and drinking water quality
- perceptions of *system risk*
- five attitudinal scales[18]
    - *trust in authorities*
    - *strength of attitudes*
    - *fair procedures*
    - *catchment risk*
    - *precautionary ethics*
- three *single attitudinal statements* that were not included in the scales
    - *technology can provide whatever treatment is needed to produce safe drinking water*
    - *when it comes to catchment protection, we should be more concerned about mining and forestry than recreation*
    - *dams currently used for recreation should not be taken away for drinking water in the future*

Seven variables emerged in the following order from the stepwise analysis and correctly classified 70% of the groups (Λ=.53; p<.001).

- Attitudinal scale: *catchment risk*
- Attitudinal scale: f*air procedures*
- Perceptions of *system risk*
- *Trust in the Water Corporation* to provide reliable information about catchment protection and drinking water quality
- Actual *knowledge* score
- Attitudinal scale: *precautionary ethics*
- Attitudinal scale: *strength of attitudes*

Examination of the means showed that the respondents who *considered the current policy to be most acceptable*:

- were more likely to be risk averse when considering the risk to drinking water from recreation
- were more likely to think that the Water Corporation used fair procedures in their decision making
- were inclined to consider that a water quality incident would be more likely, more serious and with lesser control by the authorities to stop something from going wrong with water quality if the catchments were opened up to recreation
- were more likely to trust the Water Corporation to provide reliable information on catchment protection and drinking water quality
- had a lower actual knowledge score about what's done to protect drinking water dams and catchments from pollution and contamination
- had more conservative precautionary ethics
- had slightly weaker attitudes associated with drinking water and its protection

[18] The *catchment protection* scale was not used as it represented a similar theme to the acceptance of the current policy on drinking water catchment protection.

Table A4.21 shows the discriminating variable means for each group.

Table A4.21. Means of Variables that Discriminated between Unacceptability and Acceptability of the Current Policy for the Protection of Drinking Water Catchments

Discriminating Variables	1 Most Unacceptable N=55	2 Unacceptable N=66	3 Acceptable N=139	4 Most Acceptable N=426
Catchment risk	2.4	3.2	3.4	4.0
Fair procedures	2.3	2.8	3.3	3.4
System risk	7.2	9.2	9.3	10.8
Trust in Water Corporation	2.6	2.8	3.6	3.7
Actual knowledge score	2.0	1.3	1.0	0.8
Precautionary ethics	3.6	4.0	4.1	4.4
Strength of attitudes	3.0	2.9	2.8	2.9

Similar Discriminant analyses were conducted for the next two policy options of greatest acceptance.

- *Allow off-road activities (such as trail bikes and 4 WD) in the outer catchment, but still no activities in the 2km restricted zone and no contact with water, at added cost and greater risk to your drinking water quality.*
- *Allow passive activities in the 2km restricted zone (such as walking and picnicking), but no contact with water, at added cost and greater risk to your drinking water quality.*

The acceptance groups of these two policy options were not as well discriminated as for the current policy, although still statistically significant. For both policy options, only two variables emerged in the following order and correctly classified 55% of the groups in the outer catchment option (Λ=.92; $p<.001$), and 57% of the groups in the 2km option (Λ=.85; $p<.001$).

- Attitudinal scale: *precautionary ethics*
- Perceptions of *system risk*

Examination of the means showed that the respondents who *considered each alternative policy to be most acceptable*:

- had lesser conservative precautionary ethics; and
- were inclined to consider that a water quality incident would be less likely, less serious and with some control by the authorities to stop something from going wrong with water quality if the catchments were opened up to recreation.

DEMOGRAPHIC COMPARISONS

Oneway analyses of variance were conducted to detect any statistically significant differences between responses to key questions in the survey and gender, location, age and membership of organised recreational groups. The significant differences ($p<.01$) that emerged are explained below.

Gender

- Females were significantly less likely to say they knew where their drinking water came from than males, and nominated significantly fewer sources than males (mean number of sources: females 0.8; males 1.1).
- Females had a significantly lower rating (mean= 2.3) of perceived knowledge of the protection of drinking water at dams and catchment than males (mean=3.0), and had a significantly lower actual knowledge score than males (means: females=0.7; males=1.3).
- Females were more likely to rate greater agreement with the current policy (mean = 4.4) than were males (mean=3.8).
- Females were more likely to rate greater trust (mean=3.6) in the Water Corporation to provide reliable information on catchment protection and drinking water quality than were males (mean=3.3).
- Females were more likely to rate greater trust (mean=3.8) in the Department of Health to provide reliable information on catchment protection and drinking water quality than were males (mean=3.6).
- Females had significantly higher scores on the *Recreational Enjoyment* scale (mean=3.3) than males (mean=3.0).
- Females were more likely to be concerned about *System Risk* (mean=10.6) than males (mean=9.6).
- Females were likely to hold more conservative *Precautionary Ethics* (mean=4.3) than males (mean=4.1).
- Females had significantly higher scores on the *Fair Procedures* scale (mean=3.4) than males (mean=3.2).
- Females had significantly higher scores on the *Catchment Risk* scale (mean=3.8) than males (mean=3.5).
- Females had significantly higher scores on the *Catchment Protection* scale (mean=4.0) than males (mean=3.7).
- Females considered keeping the current catchment protection policy to be *more* acceptable (mean=3.7) than males (mean=3.1).
- Females considered all other policy options to be *less* acceptable (means: 2.3; 2.5; 2.2; 2.1; 1.2) than males (means: 2.5; 2.8; 2.3; 2.3; 1.4).

Location

The eight location community sampling areas were recoded into 4 groups for analytical comparisons.

- *Far from catchments – inland* and *Far from catchments – coastal* were combined and labelled ***very far from catchments***.
- *Closer to catchments – inland* and *Closer to catchments – coastal* were combined and labelled
  far from catchments
- *Near to catchments - coastal* and *SW Near to catchments – coastal* were combined and labelled
  close to catchments
- *Near to catchments – inland* and *SW Near to catchments – inland* were combined and labelled ***very close to catchments***

The following analyses involved only the community sample.

- Those *very close to catchments* were significantly more likely to say they knew where their drinking water comes from than those living in all other locations, yet did not nominate significantly more drinking water sources than others.

- Those *very far from catchments* were significantly less likely to have ever visited or recreated at a drinking water dam or catchment in WA than those in all other locations.
- Those *very close to catchments* were more likely to rate their knowledge of drinking water catchment protection significantly more (mean=2.9) than those living in all other locations (means ranged from 2.2 to 2.5).
- Those *very close to catchments* and *far from catchments* had significantly higher scores of actual knowledge of drinking water catchment protection (means=0.9) than those *very far from catchments* (mean=0.4).
- Those *very close to catchments* were significantly more likely to rate their trust in Organised Recreation Groups to provide them with information significantly higher (mean=3.0) than those *far from catchments* (mean=2.6).
- Those *very close to catchments* had significantly lower scores on the *Catchment Protection* scale (mean=3.8) than those *far from catchments* and those *very far from catchments* (means=4.1).

Age

- Those aged 18 to 24 were significantly less likely to say they knew where their drinking water came from than those aged 40 and over, and nominated significantly few drinking water sources (mean=0.6) than those aged between 40 and 75 years (means ranged from1.2 to 1.3).
- Those aged 18 to 24 were more likely to rate their knowledge of drinking water catchment protection significantly less (mean=2.2) than those aged 40 to 55 (mean=3.1) and 56 to 65 (mean=2.9). However, those aged more than 75 were least knowledgeable with a significantly lower actual knowledge score (mean=0.5) than those aged 40 to 55 (mean=1.37), the most knowledgeable group.
- Those aged 66 to 75 years had significantly less trust in the Water Corporation to provide reliable information about catchment protection and drinking water quality (mean=3.0) than those aged 18 to 24 years (mean=3.8) who were significantly less trusting than those aged more than 75 years (mean=4.1).
- Those aged 66 to 75 years had significantly less trust in the Department of Health to provide reliable information about catchment protection and drinking water quality (mean=3.3) than those aged more than 74 years (mean=4.1).
- Those aged 66 to 75 years had significantly less trust in the Department of Water to provide reliable information about catchment protection and drinking water quality (mean=2.8) than all other age groups, and those aged 18 to 24 years were significantly more trusting than all others (mean=4.0).
- Those aged more than 75 years had a significantly higher score on the *Trust in Authorities* scale (mean=4.1) than those aged 24 to 75 years (means ranged from 3.5 to 3.7).
- Those aged 18 to 24 years had a significantly lower score on the *Strength of Attitudes* scale than those aged 40 to 55 years (mean=3.0) and 56 to 65 years (mean=3.1).
- Those aged more than 75 years had a significantly higher score on the *Fair Procedures* scale (mean=3.9) than all other age groups (means ranged from 3.0 to 3.5).
- Those aged more than 75 years were more likely to find the policy option of allowing greater access in the 2 km zone to be acceptable (mean =3.1) than those aged 18 to 65 years (means ranged from 2.5 to 2.8).

Membership of organised recreation groups

Responses from people in the community sample who advised that they were members of organised recreation groups (but not sporting clubs or groups) were compared with those of the recreator sample. Mean responses to all key questions by the community formal recreators generally mirrored those of the whole community sample and were significantly different from the mean responses of the recreators.

Community recreators:

- were significantly more in agreement with the current catchment protection policy;
- had significantly greater enjoyment from their recreational activities;
- had significantly greater *Trust in the Authorities*;
- had a significantly higher score on the *Catchment Protection* scale;
- had significantly more conservative *Precautionary Ethics;*
- had a significantly higher score on the *Fair Procedures* scale;
- had a significantly higher score on the *Catchment Risk* scale;
- were significantly more concerned about *System Risk*;
- considered keeping the current catchment protection policy to be significantly *more acceptable*;
- considered allowing passive activities in the 2km zone to be significantly *less acceptable*;

than the recreator sample.

Summary and Conclusions

It is important to note that the emphasis of this survey was on drinking water quality and its protection and the preference by different stakeholder groups for different levels of access for recreation in drinking water catchments. As such it does not deal with recreation planning *per se*. Nevertheless, it does provide important information that clearly establishes the need for drinking water catchment management to be considered in the context of meeting the future increase in demand for outdoor recreation in natural settings in the south-west of WA.

The results outlined here clearly show that the community and professional samples held similar attitudes to drinking water and catchment protection. The recreators tended to differ from the other two groups preferring to expand recreational opportunities on the catchments. Also of note was that there were differences between the recreator sample and those in the community sample who belonged to organised recreation groups (not sporting groups). Given this, it would appear that the people in the recreator sample should be considered as "champions" of recreation and looking to bettering future recreational opportunities in the light of current and future demands. Current policy was often seen as too conservative compared with practices elsewhere and their considerations of the risks involved.

Participation in recreational activities in general by all three samples was high with only 16% of the community sample saying they didn't participate. There was considerable use of *urban or city areas* for recreation by all three samples, and the community and professional recreators also made good use of the *ocean or estuaries*. While recreators from all three samples made use of *rural or bushland areas*, and *lakes or dams*, the champion recreators were more likely to participate in those locations than the others. The champion recreators were also more frequent visitors to drinking water catchments than the community recreators. However, when considering that three-quarters of the community sample had visited a drinking water catchment and that 14% of those visited *often* or *a lot*, with population growth, the future numbers of people wishing to visit drinking water catchments on a regular basis could be substantial.

Further, while it was apparent that the champion recreators found significantly more impediments to the enjoyment of their chosen activities than the community and professional recreators, there was still more than a quarter of the community recreators who were similarly discontent. When this lesser enjoyment of recreation was analysed with visitation to drinking water catchments, those who had visited or recreated at a drinking water catchment were statistically significantly less likely to be enjoying their recreational activities than those who hadn't visited a catchment. While cause and effect cannot necessarily be attributed here, as people may also be experiencing lesser enjoyment at other locations, this is an interesting finding.

Overall though, given current and potential future demands for recreation facilities and the level of current discontent in the ability to enjoy activities, it would seem that urgent attention needs to be given to recreational planning and management in the region as a whole.

There is no escaping the unequivocal agreement of the community (86%) with the current policy of drinking water catchment protection. While 57% of professionals agreed with the current policy, only a quarter of the champion recreators agreed. When respondents were given the opportunity to consider potential future policies of recreational access to drinking water catchments, again the community sample strongly supported retention of the current policy with two-thirds nominating it as the *most acceptable* option and a further 22% nominating it as *acceptable*. While a similar proportion of professionals nominated it *most acceptable*, only 10% of the champion recreators did so.

However, it should also be noted that 45% of the community and 39% of the professionals considered passive activities in the 2km restricted zone to be either *acceptable* or *most acceptable*. Seventy-nine percent of champion recreators supported this policy option.

Further 38% of the community sample and 18% of the professionals considered greater off-road access to the outer catchment to be either *acceptable* or *most acceptable*. Fifty-six percent of the champion recreators supported this policy option.

Interestingly, there was much less support by all three samples for these latter two policy options when combined to allow a considerable increase in land-based activities in the catchment.

There was very little support from the community and professional samples for either option that allowed contact with water, while 41% of champion recreators supported non-motorised water activities.

The key variables that discriminated between support and lack of support for the current policy were perceptions of catchment and system risk; perceptions of the fairness of decision making processes; trust in the Water Corporation; the degree of *actual* knowledge about catchment protection; the degree of precautionary conservatism and the strength of attitudes about drinking water and its protection.

While discrimination of support for the other two supported policy options was less certain, concerns about system risk and precautionary conservatism both emerged from the analysis.

While the champion recreators were less risk averse, less conservative in their precautionary ethics, less trusting in the Water Corporation and the fairness of its decision making procedures than the community and professional samples, they also had greater *actual* knowledge about catchment protection processes than the community sample.

The community in fact had very low *actua*l knowledge about catchment protection for drinking water and also poor knowledge about the range of sources for their drinking water. This is an important message for the authorities involved in the management of the catchments. Note should also be made of the level of trust in the authorities and organised recreational groups to provide this information. Particular attention should be paid to the quite high proportions of community and champion recreators who were uncertain about their trust in the various information providers, including the 40% of champion recreators who were uncertain about their trust in organised recreation groups. *If future informed debate is to be conducted on catchment protection, these issues of knowledge and its provision will need consideration. Note should be taken of the trusted information providers for any communication or information program, especially the trust in scientists and the Department of Health displayed by all groups.*

From the findings here it is apparent that the next, risk assessment stage of the project should initially consider the current policy option as well as the other two, somewhat supported options of passive activities in the 2km restricted zone and off-road activities in the outer catchment. All three of these policy options showed some support by at least two of the samples. While the policy option that involved passive water-based activities was supported by just under half of the champion recreators, it had little support from either of the other two groups.

The risk assessment will be conducted in the light of future population growth, potential drinking water sources and climate change. It will involve the range of professional disciplines represented in the survey, including catchment managers, recreation and tourism professionals and planners, as well as the traditional water and health managers. Key findings from this survey will be incorporated in the risk assessment, including the apparent requirement for future regional recreation planning and management, and community concerns about risk and precautionary management.

REFERENCES

Groves, R.M. (2006). Nonresponse rates and nonresponse bias in household surveys. *Public Opinion Quarterly,* 70 (5), 646-675.

Pallant, J. (2004). *SPSS Survival Manual*. Version 12, Allen and Unwin: NSW.

ADDENDUM 1
ADDITIONAL COMMENTS PROVIDED BY RECREATORS

The following comments were provided to an interviewer over the phone. Therefore, they represent notes taken of verbatim data rather than carefully scripted comments.

I think the current system of a criminal offence to be caught inside the 2km exclusion zone at the dams should be changed to a fine system instead

I believe that the water corporation have their heads in the sand, the 2km exclusion is ridiculous to other parts of the world. In some areas farm animals still get into the drinking water

I think the water should be more treated and we should have re-cycling

I believe low impact recreation bushwalking, orienteering rogaining have very little or no impact on the 2km catchment area. I strongly disagree with the lack of policing within the catchment areas, this is noticeable by bushwalking outside the 2km zone by mountain bikes, motor bikes and 4wds. I disagreed with the government not having an independent survey carried out on the water catchment to gauge damage.

I would like to say the current blanket ban to the catchment area does not work/constantly see horse riders, 4wdrivers, hunters etc around these areas close to the catchment/so obviously it is not working/increased monitoring is needed and probably would not work/would also like to say I am constantly hearing and seeing more and more signs that say you cannot do certain activities/internationally WA is getting a reputation where you can't have any fun/only things like taking photos of wildflowers etc./majority of us are able to make responsible decisions/our family is very educated and responsible

I think the main comment is the inconsistencies of applying regulations in the outer catchment/prime trail bike territory/struggle to understand that how 20kms from the dams would be affected by a few trail bikes/animals die in the water/huge trucks driving around/I questioned a Water Corp ranger and he answered, we can stop you but we can't stop the drivers, often with 12 big tyres creating havoc//causing people to engage in illegal behaviour

With the catchments native and non native animals have access to the catchments, for e.g. feral pigs which would have similar pathogens as humans. In other states and countries access to the catchments are allowed. Alcoa have mining leases throughout the catchments and process through. Spraying of plantations is also harmful to the catchments. 4wds and boats would be a big risk, if they drop fuel and oil

It has been proven that the pollution will never make it to the dam wall, there have been many tests undertaken to prove it. The area that take for the catchment is too large

I am really passionate about bushwalking, and the only areas around Perth are the catchments areas where I can do it. My recreation does not do any damage to the catchment areas, however we often see pig shooters etc that do a lot of damage

I would like to add another option. Most of the activities mentioned on selected dams

The cost of our water is too cheap, it is a precious commodity I think we take for granted. There are many other opportunities to be used in WA other than the water catchments areas

I think 2kms exclusion is arbitrary and needs to be assessed on the risk of each catchment, each catchment would have its own plan and in that plan would state the activities to be adhered to. Off trail riding, 4wds and camping should in the exclusion zone. However the water contact activities and motorised boating should be allowed depending on the assessed risk of each catchment. A lot of the statements were loaded and that is why have chosen not to answer some of the questions

Think the questions are a bit funny/not really balanced

Because of the increasing population coming into WA, the limited access needs to be reviewed for bushwalkers. There are plenty of places in the world where you can fish, boat, swim, walk right of the edge of the catchment

The questions are loaded

Walking does not cause pollution around the walking catchments. Bushwalking is healthy and the people who participate are people who like beauty and of a high class person. We are all responsible people. I don't like the fact that the one rule applies to all

I don't believe that the Water Corp is managing the water supply we have. They are not managing it responsibly. I don't believe recreation in the outer catchment areas will effect water quality

I think the questionnaire is poorly designed/most unacceptable in that the grouped options together is bad/need separate options/results will be unsatisfactory/questionnaire floored/Water Corp is a Government instrumentality and should represent whole community interests/and it doesn't do so/the WA population is growing rapidly and is there is going to be every increasing demands for recreational areas and these must be accommodated in forward planning involving recreational use of water sources/Malcolm McCusker's quote - a review of the history of this matter leaves no doubt that Water Corp gave a clear undertaking to replace the Harvey International slalom course which was lost to canoeing WA and the wider community as a result of Water Corp construction of the Harvey Dam but so far this undertaking has not been fulfilled/demonstration of Water Corps dishonesty and poor community citizenship

That the current approach to management to activities in catchments is crude. It would be much acceptable to people for better management of catchments by risk assessment. This would give better use of the assets that we have. The first step would be to compile the list of realistic potential option, this to be done independently i.e. not by Water Corp or dept of water or the recreational groups. This could be presented to government and the appropriate people would a decisions based on the independent research and options

The whole Labour government and the two water bodies that eventually closed the Logue Brook dam should be condemned/thanks to the Liberal government it is now open/governments may think twice about closing major recreational facilities

Some of the questions did not give me the opportunity to say what I want. I have specific criteria personally how I want activities to be allowed in the catchment and I think the survey stink

I think limited access to the water supply dams can be managed effectively if responsible low impact groups are targeted, licenses issued to possibly cover any additional treatment if that was necessary. I remain unconvinced it will be absolutely necessary. I think standard water treatment practices would be sufficient for low impact recreation, as I am sure there are impacts from native animals and chemical run offs at present .

I think with proper management any activity should be available/except for motorised bikes

As far as orienteering we need a close working relationship with the Water Corporation and this exists

I don't believe that there has been sufficient evidence gathered to not support water borne activities on dams etc

Quite happy with everything

Orienteering has restrictions put on us that seem ridiculous, when you have kangaroos and feral pigs excreting in the area. There are a people like trail bikers who are out there without person not doing the right thing

With water ski-ing there needs to be little control/it is currently a blanket control for all activities/other states in Australia and US allow some activities/reasonably safe and the policy needs to be looked at to allow these/other options

I support controlled to water supply catchment which include contact with the water e.g. sailing, canoeing fishing but not motorised boats. I would uncontrolled 4wds and trail bike riding they are too damaging and polluting. I would say access all water supply catchments in un-necessary, however there should be some access to some of the water supply catchments. I believe water treatment methods should be improved throughout WA. Current policy whereby Water Corp exclude access to the darling range by about 80% in not acceptable

Basically it is important to maintain a variance of opinions on drinking water dams use/need to be more flexible/some activities are less passive than others

Think management should have an open mind/users of the area should have some input

I hear stories of ridiculous restrictions and barriers in the way of benign recreation groups e.g. orienteering without rational reason for example of groups location of any toilet dug

I think overall the survey is good, however it is very directed. I think there should be more of an open forum. 4wd is much more damaging than trail bikes

Some of this information is a bit late because some of the dams have been closed or modified/the policy that we are talking about was commenced 10 years ago

I think walking and cycling should be allowed in the catchment areas, but no trail bikes or 4wds

I believe some of the questions are loaded/not enough scope to answer the way that I think/some single questions cover passive to aggressive activities/any changes listed suggest single activities would have the same impact as multiple

I believe that a limited amt of access should be allowed for passive activities, fly fishing, lure fishing and but not bait fishing that a license should be applied for and a fee paid for that license, and regular testing by an independent body of the water for a 12 months trial basis. The license fee should be $50 per annum to access the area, and possibly a limit of time to be accessed of 5am - 9pm 7 days a week. No overnight camping should be allowed in the 2km zone, due to lack of toilet facilities

I guess within the current control area passive activities excluding camping and motorised vehicles on the land and water should be allowed but these are grouped with all activities/a bit confusing to answer the way I would like

The Water Corp want to keep walkers out because we supposedly bring pathogens, what about the animals that die and excrete in the water. I wouldn't like to see 4wds and trail bikes because of the noise and the fact they just rip up the ground, I have no trouble with horses in the catchment area. I would strongly like to be able to camp overnight and draw water from the catchment

Some of the questions were virtually the same but worded differently but contradicted themselves, which made it difficult to answer. 4WDs and trail bikes do a lot of damage, mixed with alcohol and camping things get way out of hand

I think the whole issue is there is no right and wrong/negative effects of allowing recreation on the catchments/a policy where not all the catchments are affected/just allow some to be available to recreators

At the end of the day the way the population is increasing there have to be allowances for recreation /Governments have to put certain policies in place/such as allowing some recreation in the catchment areas/if properly managed there should be no problem/if all bodies work together it can be organised

Orienteering and rogaining would not cause pollution in the water catchments. The people who participate in these activities are most conscious not to pollute. I don't want any sort of motorised craft in the water or motor bikes in the catchment. When I say passive I mean on foot activities only

The word recreation is used very loosely in the survey, it can mean so many activities so it makes it hard to answer

The individual seems to be able to do what they want to do in the Darling Scarp, however the organisations seemed to be clamped down on. The management of the water catchments should be taken off Water Corp and given to DEC (CALM). Water Corp thinks they own the water. Being the events coordinator for Rogaining WA, we get all the right approvals from DEC (CALM) however we see individuals, trail bikes and 4wds do what they like without approval

We should be allowed to recreate in the 2km zone and in the water, if this requires additional treatment then so be it. Areas being able to recreate is diminishing rapidly and some of the closest areas are a day's drive from Perth

Rogaining - we need to be able to camp overnight which maybe in a catchment but not necessarily within the 2kms zone. We think the 2kms zone is too big

I think that the Water Corporation needs to have a more open mind in relation to allowing recreation on the drinking water catchment, just like they do in Canada, USA and the east coast of Australia. There is a lot evidence to substantiate that inboard motor boats do in fact more good than harm in the filtration of the water process

The drinking water catchment 2km zone is way too excessive

I think it is very hard to police off road vehicles, such as 4wds and trail bikes use the area as they wish

I think the survey should be mailed out, it is too lengthy to be done over the phone

High approval for walking running and kayaking within the 2km catchment area, less approval for mountain bikes, nil approval for motorised activities and camping

Pig shooters, trail bike riders etc are in the areas now doing damage. The questions are loaded. The questions are irrelevant because people are in areas now causing damage do something about

There are too many loaded questions! For orienteering we need the bush, the best terrain is near the water catchment. Each sport should be taken on its merit. Each sport should be risk assessed and see if there are other locations it can be done it

ADDENDUM 2

ADDITIONAL COMMENTS PROVIDED BY PROFESSIONALS

I am happy to pay increased costs for clean fresh water because it is currently grossly undervalued anyway - given its limited supply and the high demand. This exaggerates the price difference when comparing the current treatment regime with more intensive treatment. The previous survey and "community consultation" run by Water Corp and Dept of Water used misleading info and asked leading questions in an attempt to obtain responses they wanted. Hopefully the information from this survey will be used in a more balanced way.

I feel the public do not understand the potential for a water borne disease outbreak if access to the catchment and water body is increased. When I discuss this outside of the water Industry the view is that it only happens in third world countries.

To reiterate - these areas are recreation magnets. The issue is the cost to the community of dual use (recreation and water supply) in these catchments and who pays. Given increasing population these are important issues for public debate.

Protection of public drinking water supplies should be a top priority over recreation, particularly in the face of climate change, as there are many other areas to recreate that do not involve drinking water sources.

From my professional perspective, visitors, particularly from interstate and overseas, will have a very different perspective on water policy and catchment management. I believe the WA community has a relatively low level of understanding on their catchments, water sources etc and a low connection culturally to the importance of these sources. I would estimate that visitors to WA are likely to have less of a connection and understanding and therefore regard for water policy/care for catchments. I'm hoping that in your research, this market segment is explored as the strategies for management/education etc. will be different according to segment of user.

While activities such as mining and forestry have the potential to impact drinking water quality, they are governed by strict policy and money is spent in order to monitor the effects of these activities. Recreation is sporadic and much more difficult to monitor, therefore managing large scale recreation in a way that is acceptable to drinking water protection objectives is virtually impossible without huge resources. It is doubtful that these resources would ever be allocated, therefore water treatment is the only option. While I have no doubt the authorities will do everything possible to ensure contaminated water never reaches a tap, this job would be made extremely difficult by allowing known sources of pathogens into the drinking water source.

Agree with different management options outside of 2km buffer on the high waterline to within. Regulators such as the DOW Water Quality Section and Department of Health have a culture of not assisting but resisting, blocking any changes to activities or increased access to water catchments as it has no benefit to their organisation if it succeeds or consequence if it is blocked. The Water Corporation is dominated by engineers that prefer the easiest and cheapest option and using existing infrastructure rather than assisting recreation or valuing the environment. These organisations are also very political and make decisions influenced on the current political environment rather than what is best for the future.

Should any form of recreation be implemented within catchments, increased support for authorities to appropriately manage the risk of impact will be needed. Prior to increasing the exposure of risk to catchments and plantation resource, government should first look at existing recreation sites to ensure they are being appropriately managed and used before seeking additional areas.

There is plenty of areas outside of Perth's drinking water catchment areas for recreation.

Source protection of our drinking water sources is this first and most important barrier. Advanced treatment options should only be relied on if source protection cannot be effectively implemented. This does not have to be an 'us vs them' mentality. There is enough source options in WA to make all parties happy.

WA has options that other States and countries do not. We can learn from the problems/costs created by relying on treatment and protect our catchments for their drinking water value and for ecosystem benefit. Keep the public out of the 2km protection zones and have controlled only access to the outer catchment areas. There are enough catchment areas in WA to retain the split between drinking water catchments and irrigation/recreation catchments.

Excellent questionnaire..! I hope that we can keep Policy 13 in its present form and tighten regulations to ensure better enforcement and control of the activities and illegal activities that occur in the catchments.

Keep the current Statewide Policy 13 unchanged.

I am not particularly happy with some of the survey questions - they appear biased and not objective. Many of the questions and they way they are constructed do not allow the respondent to provide an accurate answer or position of their understanding. Of more concern is the preamble provided before some of the questions which is an opinion which paints a worst case scenario rather than representing the range of likely outcomes. If such a preamble is necessary then it should provide the range of opinions/views on the matter. If someone with no knowledge of the issues was presented with this information as "the facts" then it is very likely to unduly influence their responses and invalidate the survey responses. Some of the questions are ridiculous and impossible to answer eg "how much to you trust Scientists"!!!!

I think that the question "where else can you go?" should be applied to recreation applications within catchment areas. If this question is answered honestly then it will be apparent that there are enough opportunities to recreate outside of catchment areas. Those areas may require better funding and management to accommodate increasing interest in recreation outside of the metro area.

WA has enormous recreational opportunities and space outside of its water storage facilities which have a tiny footprint. It would be foolhardy to start relaxing restrictions on water resource protection.

The question about access to water catchments should be seen as part of a broader issue of community health and wellbeing, as well as cost. The current policy is driven from a point of view which sees health only as pathological. These questions (obviously) are focussed on a particular issue (about access to water catchments) and are therefore limited by it. The questions about suitability of treatment of water is also limited by reference to surface water catchments only. In the future more water will come from reuse of grey water, second class ground water and other sources, as population grows. All those options will increase costs of extraction, transport (one of the highest costs) and treatment. Conversely, surface water catchments are a finite (or declining) resource. The potential is for this to have a higher impact on alternative uses (such as public access and recreation)than on the cost and delivery of water generally.

Our lifestyle is built on a drinking water supply where we never have to question its safety. Our forefathers left us a wonderful legacy being our protected catchments. They have supplied safe drinking water for over one hundred years with minimal treatment. Such pristine supplies are rare and the envy of most others who do not enjoy such privilege. Why would we throw this away to satisfy a vocal minority who are only motivated by self interest? We are not impoverished when it comes water based recreation opportunities so our public water supplies should not be put at risk to satisfy the need of a selfish few Treatment reduces but does not remove contaminants. Why would we pay more to have a less safe water supply Doesn't make sense to me !!

High quality supply is a very valuable thing, that if anything is under-valued in WA society. One expression of this under-valuing is that water supply prices already don't take into account the full range of economic and societal costs of supply public water. In the future to become a more sustainable society WA will need to find ways to reduce our ecological footprint - our environmental impacts on resources and ecosystems that we rely on. So we should be making a high priority any measures that can ensure we can reduce rather than increase our ecological footprint. Changing policies that lead to large scale water treatment of water supplies will increase our ecological footprint, so should be avoided given a choice. And we do have a choice.

APPENDIX 5
THE HIERARCHICAL FRAMEWORK DEVELOPED FOR THE RISK-BENEFIT ASSESSMENT IN THE WA CASE STUDY

This appendix outlines the results of the development of the hierarchical framework of risks and benefits associated with implementing the three policy options that emerged from the survey phase of the WA Case study.

Current Policy

- *No public access to the 2km Reservoir Protection Zones (RPZ).*
- *Passive activities allowed in the outer catchments.*
- *Driving off-road is not allowed anywhere in the catchments.*
- *No water-based activities are allowed on the dams or in the 2km RPZ of the feeder streams.*

Alternative Policy #1

Current Policy with the addition of:

- *allow passive activities in the 2km RPZ.*

Driving off-road, and any water-based activities are still not allowed.

Alternative Policy #2

Current Policy with the addition of:

- *allow off-road activities (such as trail bikes and 4 WD) in the outer catchment.*

Still no public access to the 2km RPZ and no contact with water.

DEVELOPING THE HIERARCHICAL FRAMEWORK

As a result of individual considerations and multi-disciplinary group discussions, the two groups at the first workshop developed risks and benefits that they believed were likely with the implementation of the three policy options in the light of the hypothetical future scenario. The groups also structured the individual components into categories. Similar outcomes resulted from each group and these were then combined to provide the hierarchical framework for further consideration. This framework is shown in the tables below.

Table A5.1. Risks Associated with the Implementation of the Current Policy in the Future

Risk Category	Risk Component
Lost Opportunities	General recreational demand not met
	Nature based recreation not met
	Eco-tourism opportunities lost
	Reduced enjoyment through overload of insufficient facilities
	Lost opportunity to educate the public about catchments
	Lost opportunity for eco-education activities
Implementation & Policing	False sense of security of drinking water safety
	Overload on outer catchment and facilities
	Inability to police and enforce the policy
	It's "fun to disobey"
	Public dismissal of the policy
Cost	High cost of policing the policy
	Legal risks associated with prohibited access
	Loss of near-water land for other uses
	Inefficient use of land for water supply compared with manufactured sources
	Opportunity cost for alternative land uses

Table A5.2. Benefits Associated with the Implementation of the Current Policy in the Future

Benefit Category	Benefit Component
Public Health	Low human health risks
	Human health benefits from recreation available outside the 2km RPZ
	Provides source diversity to ensure supply under conditions of uncertainty
Water Quality	High quality water with little treatment
	Lack of reliance on advanced water treatment
	Avoids possible contaminants (prevention better than the cure)
	Reduced fire risks
	2 km RPZ filters the water
	Reduced access to feeder streams with 2km RPZ
Environmental Health	Reduced risk of dieback
	Protected ecology and maintenance of biodiversity in 2km RPZ
	Minimised erosion
	Reduced risk of introduced species and diseases
	Low carbon footprint
	Scientific benefit from intact ecosystems
Cost	More affordable water
	Low cost system implementation currently
	Public liability reduced
	Simplified management
	Convenience of maintenance of status quo
Social & Cultural Issues	Public trust in safe drinking water supplies
	Indigenous heritage protected
	Reduced recreational competition and conflict
	One policy for all in the RPZ
	Intergenerational equity of good quality water supply and environmental existence

Table A5.3. Risks Associated with the Implementation of the Policy that allows Passive Recreation in the 2km RPZ

Risk Category	Risk Component
Public Health	Increased management of health risks
Water Quality	Increased human contamination
	Need for higher level treatment
	Lack of ability for 2km RPZ to filter the water
	Increased likelihood of human contact with water
	Increased risk of fire
Environmental Health	Increased risk of dieback
	Reduced biodiversity
	Increased erosion
Cost	Increased treatment costs
	Increased management costs
	Increased policing costs
	Increased rehabilitation costs
	Inefficient use of land for water supply compared with manufactured sources
	Opportunity cost for alternative land uses
Management	Risk of "prohibited activity creep effect" (i.e. prohibited accessing water from the RPZ or motorised activities into RPZ)
	Risk of "policy creep effect" (i.e. once the policy is relaxed, it's easy to relax further with greater recreational access and alternative land uses)
	Increased complexity of management
	Increased complacency about water quality security
	Harder to police
	Increased water quality monitoring
Social Issues	Reduced public confidence in safe drinking water supplies
	Potential recreational competition and conflict
	Who pays for increased management and facilities

Table A5.4. Benefits Associated with the Implementation of the Policy that allows Passive Recreation in the 2km RPZ

Benefit Category	Benefit Component
Public Health	Health benefits from increased recreational opportunities
Opportunity	Recognises the overall value of the catchment resource
	Meets greater recreational demand
	Increased eco-tourism opportunities
	Greater accountability of actions between recreators (e.g. community "policing")
	Potential for development of improved water treatment technologies
Management	Will require better management
	Will require best practice facilities to reduce risks
	Potential to introduce fees to generate revenue for improved management and facilities
	Increased requirement for cross-agency collaboration
Social Issues	Provides wilderness and nature based experiences
	Conduit for social networking
	Increased visual amenity close to water
	Builds community support for natural experiences
Education and Research	Increased eco-education activities
	Increased opportunity for public education about catchments
	Encourage investment in catchment science and water quality research

Table A5.5. Risks Associated with the Implementation of the Policy that allows Off-Road Access in the Outer Catchment

Risk Category	Risk Component
Public health	Increased management of health risks
	Increased safety risks associated with motorised vehicles (i.e. to both passive and active recreators)
Water Quality	Increased contamination from motorised vehicles (i.e. hydrocarbons etc.)
	Increased risk of fire
	Increased erosion
	Need for increased treatment
Environmental Health	Increased spread of weeds
	Increased environmental degradation
Management	Increased complexity of management
	Increased complexity of policing (i.e. registered and non-registered vehicles)
	Difficulty in meeting created demand
Lost Opportunities	Reduced nature based recreational experience
	Eco-tourism opportunities lost
	Loss of private business opportunities with provision of free crown land access for motorised activities
Cost	Increased treatment costs
	Increased management costs
	Increased policing costs
	Increased road and track maintenance costs
	Increased rehabilitation costs
	Inefficient use of land for water supply compared with manufactured sources
	Opportunity cost for alternative land uses
Social Issues	Reduced public confidence in safe drinking water supplies
	Potential recreational competition and conflict
	Who pays for increased management and facilities

Table A5.6. Benefits Associated with the Implementation of the Policy that allows Off-Road Access in the Outer Catchment

Benefit Category	Benefit Component
Public Health	Health benefits from increased recreational opportunities
Opportunity	Recreational access for off-road vehicles
	Potential to create long distance off-road vehicle tracks
	Greater accountability of actions between recreators (e.g. community "policing")
	Potential for development of improved water treatment technologies
Management	Greater control (eg. through zoning of motorised recreation)
	Will require increased access to facilities (e.g. ablutions)
	Potential to introduce fees to generate revenue for improved management and facilities
	Recognition of what is already happening
Social Issues	Increased equity for motorised recreation
	Reduced travel costs for off-road vehicles to access facilities
Education & Research	Encourage investment in catchment science and water quality research
	Potential to educate motorised recreators

Participants commented that "off-road" and "designated tracks" needed better definition.

MAU RATING & ANALYSIS OF THE RISKS AND BENEFITS

As described in Section 5 of this report, the workshop participants were asked to rate the relative importance of the risks and benefits within their categories, and then rate the relative importance of the categories for each policy. They did this individually in the time between the two workshops.

There was considerable variation in the extremity of responses between participants. To account for this, the ratings were standardised so that all ratings within a category summed to 100 and the category ratings also summed to 100.

After the importance ratings were standardised, the component importance was multiplied by the category importance to provide an overall importance weight for both the risks and the benefits. These then were also standardised so that all the weightings summed to 100. The following tables show the standardised ratings of the risk and benefit categories, and the standardised weightings for the risk and benefit components associated with implementing three policies under the future scenario.

Table A5.7. Standardised Ratings of Risk (black) and Benefit (grey) Categories Associated with the Implementation of the Current Policy in the Future

Categories		Standardised Ratings
RISKS	LOST OPPORTUNITIES	37.89
	IMPLEMENTATION AND POLICING	36.84
	COST	25.28
BENEFITS	WATER QUALITY	27.84
	PUBLIC HEALTH	27.08
	ENVIRONMENTAL HEALTH	20.60
	SOCIAL & CULTURAL ISSUES	13.06
	COST	11.42

Table A5.8 Standardised Weightings of Risks (black) and Benefits (grey) Associated with the Implementation of the Current Policy in the Future

Risks and Benefits	Standardised Weight
Low human health risks	6.41
Inability to police and enforce the policy	5.74
Provides source diversity to ensure supply under conditions of uncertainty	4.28
High cost of policing the policy	3.97
Overload on outer catchment and facilities	3.96
Nature based recreation not met	3.95
False sense of security of drinking water safety	3.69
General recreational demand not met	3.60
High quality water with little treatment	3.60
Eco-tourism opportunities lost	3.21
Reduced enjoyment through overload of insufficient facilities	3.12
Public dismissal of the policy	2.91
Human health benefits from recreation available outside the 2km RPZ	2.85
Avoids possible contaminants (prevention better than the cure)	2.78
Lost opportunity to educate the public about catchments	2.64
Opportunity cost for alternative land uses	2.47
Lost opportunity for eco-education activities	2.42
Lack of reliance on advanced water treatment	2.29
Loss of near-water land for other uses	2.16
Reduced access to feeder streams with 2km RPZ	2.15
Legal risks associated with prohibited access	2.15
It's "fun to disobey"	2.12
Reduced risk of dieback	1.98
Protected ecology and maintenance of biodiversity in 2km RPZ	1.94
Public trust in safe drinking water supplies	1.90
Inefficient use of land for water supply compared with manufactured sources	1.89
Minimised erosion	1.84
2 km RPZ filters the water	1.79
Scientific benefit from intact ecosystems	1.73
More affordable water	1.71
Intergenerational equity of good quality water supply and environmental existence	1.65
Reduced risk of introduced species and diseases	1.52
Low cost system implementation currently	1.41
Reduced fire risks	1.31
Low carbon footprint	1.30
One policy for all in the RPZ	1.02
Indigenous heritage protected	0.99
Reduced recreational competition and conflict	0.97
Simplified management	0.93
Public liability reduced	0.92
Convenience of maintenance of status quo	0.75

Table A5.9. Standardised Ratings of Risk (black) and Benefit (grey) Categories Associated with the Implementation of the Policy that allows Passive Recreation in the 2km RPZ

Categories		Standardised Ratings
RISKS	WATER QUALITY	28.43
	PUBLIC HEALTH	28.27
	SOCIAL ISSUES	19.52
	ENVIRONMENTAL HEALTH	18.95
	MANAGEMENT	18.82
	COST	14.29
BENEFITS	PUBLIC HEALTH	25.18
	SOCIAL ISSUES	23.59
	OPPORTUNITY	21.36
	MANAGEMENT	16.16
	EDUCATION & RESEARCH	13.71

Table A5.10. Standardised Weightings of Risks (black) and Benefits (grey) Associated with the Implementation of the Policy that allows Passive Recreation in the 2km RPZ

Risks and Benefits	Standardised Weight
Increased management of health risks	4.33
Provides wilderness and nature based experiences	4.16
Reduced public confidence in safe drinking water supplies	4.09
Increased human contamination	4.08
Increased erosion	3.86
Reduced biodiversity	3.68
Builds community support for natural experiences	3.55
Need for higher level treatment	3.45
Health benefits from increased recreational opportunities	3.44
Increased likelihood of human contact with water	3.42
Potential recreational competition and conflict	3.38
Who pays for increased management and facilities	3.10
Recognises the overall value of the catchment resource	3.04
Meets greater recreational demand	2.98
Increased visual amenity close to water	2.87
Increased risk of dieback	2.73
Will require better management	2.67
Increased treatment costs	2.48
Risk of "prohibited activity creep effect" (i.e. prohibited accessing water from the RPZ or motorised activities into RPZ)	2.44
Lack of ability for 2km RPZ to filter the water	2.32
Will require best practice facilities to reduce risks	2.32
Increased complexity of management	2.31
Increased requirement for cross-agency collaboration	2.28
Risk of "policy creep effect" (i.e. once the policy is relaxed, it's easy to relax further with greater recreational access and alternative land uses)	2.24
Increased eco-tourism opportunities	2.22
Increased complacency about water quality security	2.21
Conduit for social networking	2.21

Table A5.10. cont'd. Standardised Weightings of Risks (black) and Benefits (grey) Associated with the Implementation of the Policy that allows Passive Recreation in the 2km RPZ

Risks and Benefits	Standardised Weight
Increased management costs	2.20
Increased risk of fire	2.14
Greater accountability of actions between recreators (e.g. community "policing")	1.95
Increased policing costs	1.81
Harder to police	1.77
Increased water quality monitoring	1.68
Increased rehabilitation costs	1.59
Potential to introduce fees to generate revenue for improved management and facilities	1.48
Potential for development of improved water treatment technologies	1.38
Opportunity cost for alternative land uses	1.15
Inefficient use of land for water supply compared with manufactured sources	1.00

Table A5.11. Standardised Ratings of Risk (black) and Benefit (grey) Categories Associated with the Implementation of the Policy that allows Off-Road Access in the Outer Catchment

Categories		Standardised Ratings
RISKS	WATER QUALITY	16.62
	COST	16.19
	ENVIRONMENTAL HEALTH	15.94
	MANAGEMENT	15.72
	PUBLIC HEALTH	14.52
	LOST OPPORTUNITIES	12.35
	SOCIAL ISSUES	8.67
BENEFITS	MANAGEMENT	25.37
	OPPORTUNITY	24.67
	PUBLIC HEALTH	19.01
	EDUCATION & RESEARCH	16.84
	SOCIAL ISSUES	14.10

Table A5.12. Standardised Weightings of Risks (black) and Benefits (grey) Associated with the Implementation of the Policy that allows Off-Road Access in the Outer Catchment

Risks and Benefits	Standardised Weight
Increased environmental degradation	6.01
Potential to educate motorised recreators	5.01
Increased equity for motorised recreation	4.76
Recreational access for off-road vehicles	4.41
Encourage investment in catchment science and water quality research	4.12
Increased management of health risks	3.95
Potential to create long distance off-road vehicle tracks	3.94
Increased safety risks associated with motorised vehicles (i.e. to both passive and active recreators)	3.91
Increased complexity of management	3.88
Greater control (eg. through zoning of motorised recreation)	3.78
Recognition of what is already happening	3.53
Will require increased access to facilities (e.g. ablutions)	3.34
Increased complexity of policing (i.e. registered and non-registered vehicles)	3.12
Potential to introduce fees to generate revenue for improved management and facilities	3.09
Increased contamination from motorised vehicles (i.e. hydrocarbons etc.)	2.90
Reduced nature based recreational experience	2.90
Reduced travel costs for off-road vehicles to access facilities	2.88
Increased erosion	2.69
Increased spread of weeds	2.63
Potential for development of improved water treatment technologies	2.58
Greater accountability of actions between recreators (e.g. community "policing")	2.43
Eco-tourism opportunities lost	2.36
Health benefits from increased recreational opportunities	1.96
Increased road and track maintenance costs	1.82
Need for increased treatment	1.81
Reduced public confidence in safe drinking water supplies	1.73
Potential recreational competition and conflict	1.71
Increased risk of fire	1.61
Increased management costs	1.58
Difficulty in meeting created demand	1.52
Increased rehabilitation costs	1.51
Loss of private business opportunities with provision of free crown land access for motorised activities	1.43
Increased policing costs	1.36
Increased treatment costs	1.31
Who pays for increased management and facilities	1.26
Opportunity cost for alternative land uses	0.63
Inefficient use of land for water supply compared with manufactured sources	0.56